The Engineering Solution
To
Suffering Back Pain

Colm Campbell
BE CEng MIEI EurIng

Published in 2006
by Michaela Publications

This book was typeset by Artwerk Ltd., Dublin
and printed by BetaPrint, Dublin

ISBN 0-9553445-0-6

A catalogue record for this book
is available from the British Library.

CONTENT

FOREWORD

My life changed forever twenty-five years ago, when I was taken unconscious to hospital by ambulance. After a decade of niggling back pain, the result of a rugby injury, a disc had ruptured in my lumbar spine. The pain was so excruciating I passed out.

The day before the operation I refused to have one and against medical advice and assisted by my wife I left hospital. This turned out to be the wisest decision I ever made. At the time I was unaware that "the actual proportion of all back pain patients who are surgical candidates is only about two percent." (Ref. Professor Richard Deyo of the University of Washington Medical School). This statement by an eminent professor of medicine specializing in back pain, when considered in conjunction with the claims by many orthopaedic surgeons of an eighty percent success rate, at the very least puts a huge question mark over why people readily agree to undergo invasive surgery.

The answer is, of course, ignorance. Later on in this book I will examine in detail the whole shadowy subject of invasive surgery and why "In any case spinal surgery can't really cure your back- at least not in the sense that an appendectomy can cure appendicitis. Alterations that take place in your spine during surgery may even cause or contribute to new back trouble at some time in the future." (Ref "The Back Doctor," a book by Hamilton Hall FRCS(C) an internationally recognized authority on the treatment of back pain, and a member of the International Society for the Study of Lumbar Pain).

The almost unbearable pain extended from my lumbar spine down my right leg. Terrified, unable to sit, sleep or walk, I thought my life was at its end. I had turned my back on the medical profession and, knowing absolutely nothing about back pain or its causes, wallowed in a state of self pity and excruciatingly painful terrified hopelessness until I eventually came to the conclusion that, since nobody could help me, I would help myself.

I researched the subject of back pain, and quickly discovered that sitting with bad posture is the main cause. Later on in the book I reproduce the results of a Swedish study that proves that sitting with bad posture creates greater pressure in the spinal discs than any other body position.

I reckoned that if I sat with good posture it should help. I was right. It did. I made a lumbar support for my car. It worked, giving me the incentive to perfect the product and help other back sufferers. I have been working in the area of back pain, its cause and prevention, for the last twenty-five years.

My research during this period shows that:

(1) Every spine has its own unique elongated "S" shape. Backprints are as unique as fingerprints.

(2) When that unique "S" shape is preserved there is the minimum possible stress on the spine.

(3) In engineering terms the spine is a vertical flexible column that is prone to develop a fault. The fault occurs mainly in the sitting position, when the "S" shape spine slumps forward into "C" shape. This creates huge stress in the spine with resultant back pain.

(4) When the spine is kept at all times in its own "S" shape the result, in the majority of cases, is pain free living. I am now totally pain free, but I have to work hard to achieve this by never, ever, allowing my spine to collapse from an "S" to a "C" position.

(5) The solution to a back problem cannot be handed over to a third party. You can seek help from one of the many experts working with back pain, but they are merely relieving the symptoms. Only you can eliminate the cause.

(6) You must not arrange your body around the work.

The problem was to design a product that would allow everybody to sit with their spines in the "S" position. I had made a car seat for myself that did this. However, as the length, shape and curvature of every spine is different, this posed a huge problem.

I instinctively knew I had one of the best ideas of all time. Eighty percent of the world's population suffers from back pain at some time in their lives and the possibility lay within my grasp to relieve this pain. But how could I achieve this?

I am a chartered engineer and combined my affliction with my engineering skills to design and patent a system, Spinal System-S, which provides the information to manufacture a range of office, home and car seating that guarantees perfect sitting posture. As sitting with bad posture is the main cause of back pain, sitting with perfect posture means pain free sitting. Some ten thousand back sufferers now benefiting from Spinal System-S will bear witness to this.

During my research work I read many medical papers on back pain, its causes and prevention. I also studied a large number of books on the anatomy of the spine. I found the going could be very tedious, which I suppose is a characteristic of many textbooks. My objective in writing "The Engineering Solution to Suffering Back Pain" is to explain to the reader in a simple way my approach to the relief of back pain, how the spine works, the causes of back pain, information on medical research, back operations verses non invasive approaches and why there is no such thing as a quick fix for pain caused by a mechanical malfunction.

I took responsibility for my own life, researched the causes of back pain and applied them to invent and patent a system that guarantees perfect sitting posture. Of course a huge amount of pain is due to reasons other than sitting with bad posture.

I then researched how the majority of the world's population abuse their spines on a continuous basis when not sitting. Using this information I designed a system, Spinal System-Life, which lays down three principles that provide pain free living. I explain Spinal

System-Life in detail in the following pages. It has resulted in me living a pain free active life.

Had I gone under the knife twenty-five years ago I could now be like the many unfortunate examples of botched operations I meet on a regular basis.

CHAPTER ONE

RUPTURED DISC

"The wisest decision I ever made"

"You're wanted on the phone."

I stood up from my desk to take the receiver on a nearby table and hit the floor unconscious. My fifth lumbar disc had ruptured, had shot out of its housing like toothpaste out of a tube. And, like toothpaste, could never be put back in. From that moment on my life changed. Forever.

I awoke in an ambulance taking me across the city to hospital. I became aware of the siren ringing and flashing lights being reflected through the windows. As I regained full consciousness I could hear a mumbled conversation between the ambulance crew. I got the gist immediately.

"It's not my heart. My back." The effort of saying this caused fresh shock waves of excruciating pain, low down to the right of my lumbar spine. I gritted my teeth, desperately trying to stifle the scream of agony. Holding my teeth together, mouth closed, caused me to stop breathing. I was winning, the embryonic scream stillborn. However I needed oxygen, urgently. Gulping for air caused the pain to shoot to an even higher level of intensity. I've no personal experience, but a red-hot needle puncturing my spine would have been a mere pinprick in comparison. Teetering on the verge of unconsciousness the ambulance hit a pothole and once more I passed out.

When the ambulance arrived outside the Accident and Emergency Department I was fully conscious and even now, some twenty-five years later, am appalled and angry at what then took place. I was lifted bodily by two of the ambulance crew and dumped into a wheelchair. Again I passed out. I still find it incredible that people working in the medical field could be so callous. Everybody knows, from their schooldays, that sitting with bad posture is the main cause of back pain, which is the second most common pain after headache. As eighty percent of the world's population suffers from back pain at some time in their lives (Ref. The World Health Organisation) the ambulance crew must have been aware that the worst thing you can do with somebody in my condition is to drop them in the seated position. Further on I will write about a Swedish study which gives the pressure in the third lumbar disc when a person is lying, standing, bending forward and sitting with bad posture. The latter position causes more pressure than any other position.

What happened next was potentially worse than the ambulance crew depositing me into the wheelchair. I was awoken the next morning by a gruff, irritated, voice ordering me out of bed. I lay doggo, terrified to move. It wasn't the voice that frightened me, but the thought that even a slight body movement could increase the pain.

"Get out of bed!" He spat out these four words with venom, wasn't used to being disobeyed.

It took me some considerable time to do his bidding, each tiny movement causing shock waves of agony. I am pretty fiery by nature, have always put bullies in their place, but at that moment my resistance was at its lowest. Pain was a big factor in causing this; the biggest by far was fear. Fear of the unknown. I had never experienced such debilitating pain. The sheer terror of not knowing where it came from, what caused it, would it be with me for the rest of my life, was this the end of life, made me totally subservient to this insensitive tyrant. Oh yes. All smiles and nudge-nudge jokes with the matron and senior nursing staff, he seemed to have a pathological hatred for patients. His bedside manner was one of intimidation- shut up, don't ask any questions and do what I say.

I now half stood, half leaned on the bed; my back locked into a bent position - a broken man, needing compassion and a few kind words like "I know you are in pain and frightened. But don't worry. Most back pain, even chronic, will disappear in time." (I make a point of saying something like this to most back sufferers I meet.) But no. He came in with: "Bend, touch your toes!"

I looked up at him from my half bent position. I could see a white beard below an angry face and knew he was the "voice." Surgeons always dress in sports jackets, student doctors in white shop coats. His was a rumpled red and green check affair, the five students in freshly starched outfits, each holding a pen in one hand, a clipboard in the other, with the insignia of office, the stethoscope, hanging from the neck.

Of all the things you can say to a chronic back sufferer this is by far the worst. It had the same effect on me as being hit on the chin by a right hook from a professional heavyweight boxer. I went out like a light, my body totally incapable of handling the psychological reaction to being ordered to do something guaranteed to cause an enormous increase in pain.

The day before the operation I decided not to have one. This was the greatest, wisest, decision I ever made. In twenty-five years I have spoken to somewhere in the region of four thousand people who underwent surgery. The number of successful operations is less than one percent. I define a successful operation as one that restores the body to the condition that existed before the outbreak of pain. I have a link on my website www.back-shop.com to the back pain page of the famous Mayo Medical Clinic. They state, "Before you decide on back surgery consider getting a second opinion."

Unlike many operations whose success rates are fairly accurately known to the general public, the whole subject of back surgery is shrouded in uncertainty. What is your perception? Ask a number of people to give you what they perceive is the percentage success rate for spinal surgery. You will find a huge variation. I have been doing this for twenty-five years.

My findings are,

- 10% say zero
- 10% say five percent
- 50% say less than twenty percent,
- 20% say twenty to fifty percent,
- 10% say over eighty percent.

Professor Deyo of the University of Washington in an article published in August 1998 states, "The actual proportion of all back pain patients who are surgical candidates is only about 2 percent." So you realize why I consider myself lucky to have refused surgery.

The Mayo Clinic also contends "Long term outcomes also are often similar following less invasive treatments." While this is a big step forward for "medicine mankind." I would disagree. I applied my own non-invasive treatment and am now completely pain free. Had I had an operation this certainly would not be the case.

My decision not to have an operation was not based on any clever decision on my part. I knew absolutely nothing about the spine and I certainly did not know that "even for patients with a slipped (herniated, ruptured, or prolapsed) disc spontaneous recovery is the rule." (Ref. Scientific American, August 1998) The decision was arrived at using a mixture of native cunning, being dropped into a wheelchair and ordered to bend and touch my toes. The insensitivity of the orthopaedic surgeon, made me suspect the competency of the medical people and sheer unadulterated fear. The person in the bed beside me was a member of the traffic department of the Garda Siochana (Police). He too was awaiting an operation on his back and was in great pain. I vividly remember his painful moans, which were continuous and prevented me from sleeping. I presume we must have been in a queue awaiting an operation. Luckily he was ahead of me. After the operation he was in a worse state. The decibel level of the moans had greatly increased and the duration between each moan decreased in the same proportion.

I instantly made up my mind and couldn't wait for my wife Chris to come on her nightly visit. When she arrived events happened with astonishing speed.

I said, "I'm not having an operation!" She immediately came back with, "Let's go!" She got my clothes from a locker, helped me into them and within ten minutes the pair of us stumbled and swayed out of the hospital entrance, my right arm around her shoulder and she taking most of my weight, almost buckling under my thirteen stones. We had no communication with the hospital authorities, either during our flight or afterwards.

The initial relief of escaping from that inhumane hospital ward helped me endure the agonizing car journey. Throughout my seven days residence none of the staff ever spoke a consoling word, gave any hope for the future, or information about back pain. Once home, lying on the dining room floor, the enormity of my predicament hit me and I became deeply depressed. I had been thinking only about the pain and myself. But if I was to be incapacitated for life what was to become of my wife, Chris and family? We had four children, Paki the eldest at ten, Sean, Claire and Gordon. Tara wasn't born at the time.

I had turned my back on the medical profession, so couldn't go back to them for help. After my recent experience this was out of the question anyway.

After several weeks the pain eased and I tried walking a number of steps at a time. Getting from a lying to a standing position took ages with each tentative upward movement creating fresh painful assaults on top of what I was already enduring. But I made it. Then I slowly inched forward, body as tense as a soldier walking across a minefield. I was lucky I suppose. I wasn't waiting for a large explosion, just more pain. And like the soldier who successfully gets through without the big bang, I had the same feeling of joyous relief. There was no increase in pain and slowly, over the period of a week, I was able to gradually increase the distance I could walk.

But I still couldn't sit. I instinctively knew this was the core of my problem. I experimented by jamming a rolled up towel into my lower back as I tentatively sat on a dining room chair. Anything else was too low. I had experimented using a towel while flat on my back in bed and it helped. It had some success in the chair also. After six weeks I was a great deal better, but still couldn't sit for more than twenty minutes, even with the help of the rolled up towel.

I decided to go back to work, a journey of about thirty minutes. I managed to lever myself into the car seat, complete with rolled up towel, started the engine, looked into the rearview mirror, pressed down the clutch pedal to engage reverse and bang, the pain was back with a vengeance. It took Chris nearly thirty minutes to extricate me from my locked position behind the steering wheel.

In deep depression, I was back to where I was six weeks ago. But was I? I had come a long way since that ambulance journey. I was progressing ok until I used the clutch. I instantly remembered an incident on O'Connell Bridge some years previously. I was stopped at traffic lights, the light turned green, I pressed down on the clutch and almost landed in the back seat. Luckily I had the presence of mind to hang on to the steering wheel. My car was new, the driver's seat wasn't securely bolted to the floor and the action of pushing down on the clutch pedal resulted in a huge force traveling up my leg into my lumbar spine and pushing against the back of the seat with sufficient force to undo the floor bolts. This force was the cause of my recent setback.

I considered the problem from an engineering point of view. I knew my body was telling me that sitting was the cause. I could just about manage to lie, walk and stand. Sitting caused pain. I considered the spine as a vertical flexible column, which is prone to develop a fault; it collapses from an elongated "S" shape to a forward slumped "C" in the seated position. A basic engineering principal states that bending a stiff object creates stress in the object, the greater the bend the greater the stress. Hold a wooden ruler down low in your left hand, grasp the top with your right and bend it. Releasing your top hand releases the stress and the ruler whips back to its original position so fast the eye cannot follow. I knew instinctively this fact was the cause of my back pain. The extra force coming up my leg, resulting from the action of using the clutch had caused the abandonment of my car journey.

I reckoned that if "C" (slumped forward in bad posture) caused back pain, then sitting with perfect posture, which I call the "S" position, should both relieve and prevent pain. I designed a crude instrument to measure the shape of the lumbar spine. I sat with my spine in what I considered to be its perfect position and got Chris to draw the profile on a piece of cardboard. I then made a template from the cardboard and, using a hacksaw blade and some polyurethane foam, made a lumbar support with a profile exactly matching the template. Some three days after the first aborted journey I put the lumbar support into my car and drove to work. A bit tender, yes, but no unbearable pain. A week later I was able to drive pain free for up to two hours.

Now some twenty-five years later and using my infinitely more sophisticated Car Seat Mould, I can drive for at least five or six hours at a time totally pain free. I have driven these distances several times in France over the last couple of years.

The result of my research into the cause of back pain is my patented Spinal System-S, which is unique worldwide, guarantees perfect posture and over the last twenty five years has given pain free sitting at work, in the home and the car, to something in the region of ten thousand people.

However sitting is only one of the many activities of daily living. I studied my own lifestyle and realized that, apart from sitting, I was abusing my spine on a constant basis throughout the day and night. I rectified this, resulting in my spine never moving out of its own "S" shape and now don't suffer pain.

***I call the method I use for pain free living
"Spinal System-Life." (This incorporates "Spinal System-S")***

My intention on the following pages is to lead you through this process of research very much as I did, so that at the end of the book you will have a good deal of knowledge of the spine, how back pain is caused and how it is prevented.

CHAPTER TWO

THE SPINE

"S = Perfect Posture"

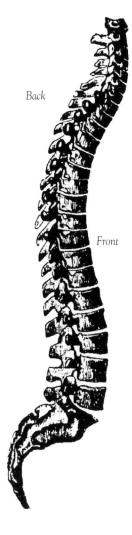

Back

Front

When I started researching back pain, its causes and prevention I had like most people a limited knowledge of the spine. I knew it was made up from cylindrical shaped bones, called vertebrae, stacked one on top of the other with the small bones at the top gradually increasing in size down the length of the column. I also knew that between each pair of vertebrae were soft pads called discs. I knew there was something called a spinal canal running the length of the spinal column behind the vertebrae and through this canal runs the spinal cord enclosing bundles of nerves that connects the brain to the various parts of the body. And that was about it.

I knew that sitting with the spine bent in the forward "C" position was the cause of my pain. To even begin to discover why this is true would mean that it was necessary to obtain some rudimentary knowledge of the spine.

I studied medical textbooks and learned that the spine, as shown in the drawing, is made from twenty-four bones called vertebrae stacked on top of each other, with a flexible pad, in my opinion confusingly called a disc, between each pair of vertebrae. The term "disc" causes a great deal of confusion and leads many people to think that the word "disc" refers to a one piece flexible pad. In fact a disc is only one of three items that make up the flexible connector between every two vertebrae. To understand how the spine works, what causes back pain and how this pain can be avoided and relieved, it is essential you realize this fact. I will talk in detail about this later. The vertebrae are tightly attached to the so-called discs.

Two long ligaments (Bands of tough flexible fibrous connective tissue linking bones together.), one running down the front of the stacked column and one running down the back of the column, bind the vertebrae together. These are not shown.

The spinal column acts as a conduit called the spinal canal which conveys the bundle of nerves called the spinal cord from the brain to the various parts of the body. Between each pair of vertebrae two nerves known as nerve roots emerge from the spinal cord, one on either side of the vertebrae. They supply the left and right sides of the body.

My fifth lumbar disc, the lowest disc, had ruptured, made contact with the nerve root resulting in an almost unimaginable high level of pain down low in the right side of my back This was accompanied by pain running down my right leg known as sciatic pain, which if anything was probably even more excruciating.

It was now some time since I had left hospital. During this period I constantly monitored my body movements twenty-four hours a day, kept note of everything that caused pain and then worked out a method to avoid repeating the pain causing movement. I was coping quite well with day-to-day life by always keeping my spine in its own unique "S" shape and never ever sat unless my spine was supported with a special lumbar support cushion which I had designed.

I kept my custom tailored car seat mould permanently in my car. The dreaded forward "C" position was never attained. To do so, even for a fleeting moment, could result in fresh assaults of pain. The perennial problem was how to put on socks and shoes.

Lifting was totally avoided. If something had to be lifted, I had no hesitation in asking anybody, even a stranger, to do this for me. I never cease to be surprised by just how helpful people are when they realize you are a back sufferer. As over eighty percent of people suffer back pain at some time in their lives, everybody either knows a sufferer or is one themselves and so are sympathetic to the affliction.

In spite of what the medical profession was advising I realized that an orthopaedic mattress should be avoided. Lying on a hard mattress, whether on your side or back, does not support the spine, allowing it to sag. I knew that supporting my spine when seated relieved pain. Logic told me that supporting the spine in bed was imperative. I used a lightly rolled up towel placed across the bed to support my lumbar spine. Whenever I found a position that gave relief it would last only for a short time before the outbreak of fresh pain forced me to seek a different position. I achieved this by bearing down on the mattress with one elbow and using this as a fulcrum, rotating my body around the elbow with the palm of my other hand. This put the minimum pressure on my spine.

Therefore, as I continued my research into the working of the spine with the ultimate objective of designing a system to relieve pain, I was relatively pain free.

I had been told by the hospital that a "prolapsed disc" had caused my pain. I had no idea what this meant. To be truthful, I didn't want to know. I was paralyzed by a two-pronged attack on my senses, one mental and the other physical. To say I was frightened was an understatement. There was no hope; my life was at an end. What was to become of my family? Not once did anybody in the hospital tell me that most back pain disappears in a matter of weeks and that ninety percent of slipped (prolapsed, ruptured, or herniated) discs rectify themselves in time. This simple piece of information would have given me hope, relieved my feelings of utter despair and helped me to cope with the bouts of agonizing pain that, night and day, assaulted my body. It has been my experience over the last twenty-five years that most back sufferers whose pain reaches excruciating levels will agree to any medical invasive treatment that gives even a glimmer of hope of eliminating their pain. Sheer unadulterated fear made me differ from the norm.

Having studied the subject for some considerable time I knew a great deal about the anatomy of the spine. This was in theory. I needed a more hands-on knowledge. I obtained a full sized plastic model of a spine in a medical supplies shop that fulfilled this requirement to some extent. I could bend it into the "S" and "C" shape and observe what happened. While the plastic "disc" was much softer than the rigid "vertebrae" there was a complete absence of a bulge when the spine was bent forward.

I set up a simple experiment to see what happens when the spine goes out of its natural "S" position into a leaning forward "C" position and why I got referred pain down my leg.

I define the "S" position of the spine as "the position of the spine that results in the minimum possible pressure in the intervertebral discs."

S = Perfect Posture

The experiment was to simulate the spine under workshop conditions. Then bend the spine and observe what happens. I chose the five vertebrae of the lumbar spine because this is where most back problems occur.

I obtained five wooden children's building blocks, representing the five vertebrae of the lumbar spine. They measured 50 mm x 50 mm x 50 mm

From an old inner tube of a bicycle tyre I cut two strips measuring 30 m.m. wide and approximately 300 mm long. These were to represent the two longitudinal ligaments that run down the front and back of the spine and are rigidly attached to the vertebrae. I got a slab of jelly, the kind you dissolve in boiling water to make jelly for a trifle. This

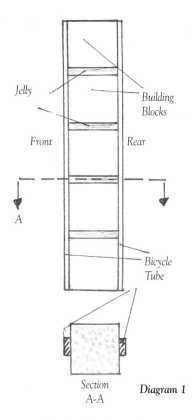

Jelly

Building
Blocks

Front   Rear

A

Bicycle
Tube

Section   **Diagram 1**
A-A

was to represent the consistency of the intervertebral disc structure. (I knew this has a soft inner core somewhat like the consistency of jelly.)

The soft inner core of my fifth lumbar disc had shot out of its housing and once out cannot be put back. It's known as a slipped, prolapsed, ruptured, or herniated, disc. I will write in detail about the intervertebral disc structure later.)

The slab of jelly was 12 m.m. thick. I cut it into pieces measuring 50 m.m. x 50 m.m. and placed one between each building block. I then screwed the strips of rubber tube, under tension, on to the two opposing sides of the 5 building blocks.

I now had a fairly rigid, crude, mock-up of a spine. (See diagrams.) Running up either side of the building block column can be seen the two strips of bicycle inner tube representing the front and rear longitudinal ligaments. These are screwed into the blocks. The screws are not shown.

Diagram 1 represents the lumbar spine with no bearing weight. Section A-A shows the jelly level with the sides of the building block. The bicycle tube can be seen in cross-section on either side of the block.

Diagram 2 shows the "spine" bent forward in the "C" position. Like stress produced in a wooden ruler as a result of bending, I knew this would create a similar effect and wanted to obverse it. Section A-A

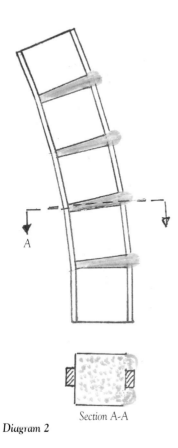

A

Section A-A

Diagram 2

shows the jelly bulging out in two places on the right hand side of the building block. The rubber strip prevents it from bulging directly backwards and this is exactly what happens in a real spine. Every spine has two longitudinal flexible ligaments, something like my two rubber strips, running from the cervical to lumbar spine. The rear (posterior) ligament prevents damage to the spinal cord that runs in the spinal canal at the back of the spinal column.

Unlike my experiment, where the jelly representing a disc covers the complete area of two wooden adjoining building blocks representing the vertebrae, the soft jelly-like material of the disc in the spine is surrounded by a casing made from onion-like layers of flexible fibrous material, called the annular ligament that also binds the two vertebrae together.

Looked at in cross section it is something like a cross section of a tree, with a blank area in the middle representing the disc.

The areas of the vertebrae in contact with the disc have highly polished surfaces called cartilage plates. Their function is to provide sustenance to the disc.

The disc, annular ligament and cartilage plates are collectively known as the disc complex. It is extremely important that you commit this to memory. Many people think the disc is a semi-hard disc shaped object that slips out of position and can be "put back in."

A prolapsed (slipped, herniated, ruptured) disc is the result of a tear in the outer wall of the annular ligament, resulting in the pulpy mass of the disc bursting out through the tear with the possibility of making contact with the nearby nerve leaving the spinal cord.

My fifth lumbar disc had ruptured and made contact with the sciatic nerve, which exits the spinal cord at this point and goes through the buttocks and down the leg, resulting in excruciating pain in both my lumbar spine and my right leg.

The drawing shows a spinal column. There is a total of 24 vertebrae, not including five fused bodies that make up the sacrum and the 4 fused bodies that make up the coccyx, or tail bone.

The vertebrae, as you can see, are far more complex than the ones I have drawn to show my simple experiment. I will go into more detail later. For the moment just bear in mind that the bony pieces sticking out at the back of each vertebra are an integral part of the

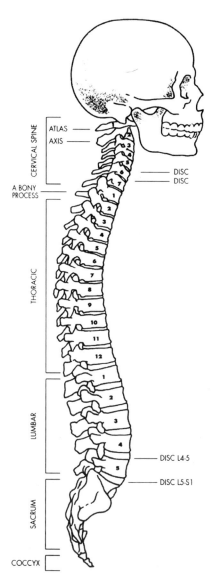

vertebrae. They are inclined more or less downward and are designed to protect the spinal cord.

- The neck or cervical area has 7 vertebrae. Numbered from C1 to C7 downwards

- The thoracic area has 12 vertebrae. Numbered fromT1 to T12 downwards.

- The lumbar area has 5 vertebrae. Numbered from L1 to L5 downwards.

- The sacrum is a bony mass made up from 5 fused vertebrae.

- The coccyx, or tail, made up from 4 small vertebrae.

- The atlas is the name given to C1 the first cervical vertebra in which the head rests.

- The axis is the name given to C2.

Between each vertebra is the disc complex. Its function is to permit movement of the spinal column and also acts as a cushion between the vertebrae to enable the spinal column to withstand the normal jars and jolts of everyday living. There are a total of 24 disc complexes.

Armed with this information and having no access to specimen spines, I considered the problem of how to witness what happens when the spine sags into a forward "C" position. I then set up the experiment just mentioned.

CHAPTER THREE

HOW THE SPINE WORKS

"In engineering terms the spine is a vertical flexible column"

During the last twenty-five years I have had approximately ten thousand customers. On top of this I give in-house lectures to company employees on back pain its cause and prevention and talk on a one-to-one basis to an average of thirty back sufferers per week. All this would add up to something in the region of forty thousand people suffering from what is the world's second most common pain, after headache.

The one thing most of these people have in common is an extremely sketchy knowledge of the spine. They all know that sitting with bad posture is the cause of most back pain; it has been drummed into them from an early age by parents and teachers. When asked why bad posture causes pain, their blank faces give the answer. They have accepted it as a fact of life, never asking why. My simple experiment gives one of the reasons why. There are many more. It is important if you are a back sufferer or don't wish to become one that you have a clear knowledge of the spine. I have read many books on the subject and found many of them confusing. I am now going to explain the structure of the spine and kill off some of the myths associated with it.

It is my absolute conviction that unless you have a good knowledge of the spine, its structure, how it works, how faults occur, how pain is caused and how it can be prevented, you are unlikely to live a pain free existence. In order to emphasize the importance of some points, on several occasions in the following pages I will be somewhat repetitive.

In engineering terms the spine is a vertical, flexible, column that tends to develop a fault. Under certain circumstances it changes shape from an elongated "S" to a forward "C", resulting in great stress from the neck (cervical) through the upper back (thoracic), to the lower back (lumbar) where it is greatest.

Hold the end of a cheap ballpoint pen in your hand. With your other hand bend it from the top. Continue bending and it will break, down low where you are holding it, where the stress is greatest. In the pen's lumbar region, so to speak. The same forces are at work when the spine bends forward into a "C" position.

I have measured the "S" shape, (perfect posture when the intervertebral disc pressure is the lowest possible), of ten thousand spines and have discovered that the curvature, length, width and physical make up of the bony structure, etcetera, are unique to the individual. I have coined the expression,:

"Backprints are as unique as Fingerprints."

THE SPINAL COLUMN

The drawing shows the same right hand side view of the spine as in the previous chapter. As you can see the spine is an elongated "S." The bony vertebrae, like the wooden building blocks in my experiment, are stacked on top of each other. They are kept apart by flexible ligaments known as "discs." I used household jelly to represent them in my experiment.

A ligament is "a short band of tough fibrous tissue which connects two bones or holds together a joint." In my opinion calling the flexible ligament that binds two vertebrae together a "disc" causes great confusion among the general public. The common meaning of the word "disc" is a flat thin round object. (I know in the Oxford English Dictionary another meaning of the word is "a layer of cartilage separating vertebrae in the spine," but how many people are aware of this?) And so in my experience the majority of people think the vertebrae are kept apart by firm somewhat flexible pads that can "slip" backwards out of position, with resultant pain. Hence the expression "slipped disc."

I will always refer to the three components that separate any two vertebrae as the disc complex and will shortly go into more detail.

Unlike my building blocks, which are all the same sizes, vertebrae gradually become bigger from the top downwards. They also have extra bony structures sloping downwards at the rear called vertebral arches, each containing a hole through which the spinal cord travels on its journey from the brain, down the back of the spine, to the various parts of the body.

Each vertebral arch has two upward pointing pieces of bone and two downward pointing pieces of bone. These lightly lock into corresponding pieces of bone in the vertebra on top and the vertebra below to form what are known as facet joints.

Each vertebral arch has two bones sticking out sideways, one to the left the other to the right. These are called the transverse processes. (A process is an outgrowth in a organism.)

Each vertebral arch has a wide thin bone at the back that is directed backwards and downwards. This is called the posterior (rear) process.

As already mentioned the complete column rests on the sacrum, a triangular bone formed from five fused vertebrae and situated between the two hipbones of the pelvis

At the bottom of the sacrum is the coccyx or tailbone, formed from four more fused vertebrae.

As we have learned already, running down the length of the spinal column, one at the front and one at the rear, are the longitudinal ligaments. Just as the disc complex binds the vertebrae horizontally, the longitudinal ligaments bind them vertically.

VERTEBRA

I sketched the drawing below from an actual vertebra of the lumbar spine.

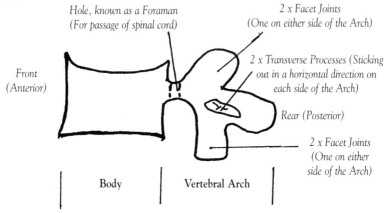

The body is roughly cylindrical in shape with sides curving in slightly as shown. The body's fairly smooth surface encloses an interior that consists of a honeycomb of bone. The top and bottom surfaces of the body are covered by highly polished surfaces called Hyaline Cartilage Plates. The cartilage they are made from resembles the surface of a knuckle of chicken bone and the disc, which I will talk about presently, is supplied with nutrient through contact with the cartilage plates,

The Vertebral Arch as can be seen from the drawing that shows a side view and the following plan view, is an extension of the body at its rear.

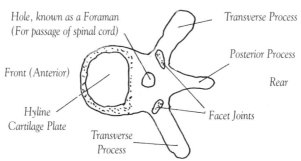

The Arch has a hole through which the spinal cord travels on its journey from the brain to the lumbar spine. It has seven bony outgrowths known as processes.

There are the four **Facet Joints**, two of which can be seen on the side view drawing. The other two are on the far side of the arch. As they are more or less in the vertical, only a cross section of each joint can be seen on the plan view. The other two are on the bottom of the arch. Two facet joints lock into corresponding joints on the vertebra above and two on the vertebra below. The bearing surfaces where they meet are covered in low friction slippery glasslike cartilage. The function of the facet joints is to guide the spine up and down in the vertical direction and to limit rotation of the spine.

The Posterior Process is directed downwards and backwards It functions as an anchor to which muscles and ligaments are attached and also protects the spinal cord, where it leaves the spinal canal.

The TwoTransverse Processes stick out sideways and also act as anchors for   muscles and ligaments.

MECHANICAL FUNCTIONS OF THE SPINE

It provides a casing, the spinal canal, to protect the spinal cord that runs from the brain down to the lumbar region. It also protects the nerve roots that emerge from the cord at each junction between the vertebrae and then extend to the various parts of the body.

It is designed to keep the body upright to permit great flexibility and is actuated by the motor functions of the body. Movement is allowed by means of the flexible disc complex between the vertebrae and the facet joints.

It is a structure designed to anchor the powerful muscles of the trunk, shoulders and neck.

THE DISC COMPLEX

Between the vertebrae is the disc complex. In my opinion an understanding of its structure, function and how it develops faults, is critical to the elimination of a great deal of back pain and its prevention. Having this knowledge and putting it into practice on a continuous basis throughout the day will, in most cases, result in pain free living. I have been totally pain free for many years. However this did not come about automatically. I had to work hard at it over a period of time in order never to forget to keep my spine at all times in its "S" shape.

The Disc Complex is the collective name for the two Hyaline Cartilage Plates, the Disc and the Annular Ligament.

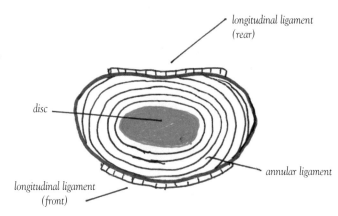

longitudinal ligament (rear)

disc

annular ligament

longitudinal ligament (front)

The drawing shows a cross section of the disc complex. The longitudinal ligaments that bind the front and rear of the vertebrae together are also indicated. The hyaline cartilage plates, one in contact with the top of the disc, the other in contact with the bottom of the disc, are not shown.

The Hyaline Cartilage Plates are glasslike plates deeply embedded into the upper and lower surfaces of the bodies of every vertebra. Their highly polished surfaces provide nutrient to the discs by diffusion. The discs are not attached to the plates and are able to move around the surfaces. However, once a disc moves out of contact with the plates, they lose their supply of nutrient, die and waste away.

The Disc (Nucleus Pulposus) is the dark area shown in the centre of the drawing. It is composed from a soft jelly-like material. When young the moisture content of the disc is approximately 90%. With aging the moisture content gradually decreases. This process can commence in the twenties.

I have already mentioned that many people think that the disc is a semi rigid disc or plate that separates the vertebrae and is liable to "slip" backwards out of contact of the vertebrae. They tell me the treatment they receive results in the "slipped disc" being pushed back into position. Some medical books that use the expressions herniated, ruptured, discs compound this confusion.

It's not the disc that ruptures, but the casing
(annular ligament) that surrounds it.

When the annular ligament ruptures the enclosed disc, which is always pressurized, shoots out through the tear in the outer wall of the annular ligament like toothpaste out of a tube. And like toothpaste cannot be pushed back in.

The disc acts something like a damper on a motorcar suspension and helps prevent the spring, in this case the annular ligament, bottoming out. The disc is not essential to a person's anatomy. Alexander Walker-Naddell FRCS, consultant orthopaedic and neurosurgeon, was able to demonstrate in the pathology department "that many of the specimens who had died of various diseases had no discs at all, yet during their lifetime had had no history of back trouble whatsoever." (Reference to Walker-Naddell FRCS to follow in Chapter Ten.)

The Annular Ligament is elastic, compressible and closely bonded to the upper and lower vertebrae. It is composed of fibrous tissue and laid out in layers (laminates) that completely surround a soft jelly-like core, the disc. The fibres that make up the layers are arranged in parallel lines, as can be plainly seen in my drawing. The fibres in adjacent layers are arranged obliquely in opposite directions, thereby greatly increasing the strength of the annular ligament. Just like radial car tyres.

The annular ligament allows movement of the spinal column, dampens out jolts associated with day-to-day living and is really a combined coil spring / ball joint. For example when the foot hits the ground when running the entire 24 disc complexes are compressed, some to a greater extent than others. A split second later when the foot leaves the ground the compression is released and the disc complexes literally spring back to their original positions. The other foot hitting the ground repeats the cycle and so on. It is also possible to rotate the spine left or right while running. (Something like a ball joint.)

The schematic drawings below show the disc complex as it distorts from a position when a person is standing to when the person is still standing, but holding a weight, to when the spine is bending forward.

The first drawing shows the disc complex without any weight bearing, as in standing. The edges of the annular ligament can be seen flush with the sides of the vertebrae.

The egg shaped disc can be seen wrapped up in the layers of the annular ligament. The vertical line running through the disc shows it to be symmetrical.

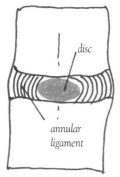
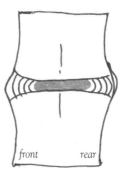

The second drawing shows the disc complex when a person is standing and holding a heavy weight, with the spine in the S position. You can see plainly that the edges of the annular ligament have been squeezed so that they are protruding over the edges of the vertebrae. The distance between the vertebrae is now less than it was before the weight was applied, as a consequence of the weight. The thickness of the disc has also been reduced and it has been stretched horizontally.

The third drawing shows the spine bending forward as in touching your toes or working at a desk. As you can see the annular ligament is compressed at the front and stretched at the rear. Simultaneously the disc is squeezed backwards due to the vertebrae on top and below coming closer at the front and wider at the rear. It is now almost completely to the right of the vertical line.

THE BODY'S SUSPENSION SYSTEM

To help you get a better idea of how the disc complex works compare it with the mechanism below that provides the suspension in a motorcar. This has an open coiled spring that gives a smooth ride by absorbing the shock forces generated by potholes and uneven road surfaces.

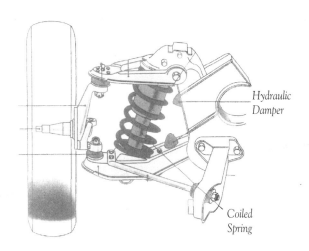

Hydraulic
Damper

Coiled
Spring

When the car hits a deep pothole the sudden force could cause metal-to-metal contact between the coils of the spring. Many drivers will have experienced this "bottoming out" with resultant loud bang after driving over a particularly deep hole. To prevent this happening the spring has an hydraulic damper attached that provides a cushioning effect, prevents metal-to-metal contact and stops the spring oscillating. (The annular ligament can rotate somewhat, the facet joints restrain it. The coiled spring does not rotate.)

The action of the disc is similar to the hydraulic damper. The disc prevents the annular ligament " bottoming out" just as the hydraulic damper prevents the spring doing likewise.

Both the jelly-like material in the disc and the hydraulic oil in the damper are always pressurized. Also the annular ligament can function without the disc being present, just as the coiled spring can work without the hydraulic damper.

A faulty damper can be replaced, a disc cannot.

A faulty hydraulic damper, or the complete removal of the damper, does not affect the performance of the car, provided it is driven with care. If fast driving over uneven pot holed surfaces, tight cornering at high speed, heavy braking and general misuse are avoided you would not be aware the damper is missing or not working.

The same applies to your body. Should the disc be absent from your disc complex due to, for example, a severe rupture you would be able to carry on life as usual and be totally unaware of this fact.

The following quotation by Walker-Naddell FRCS from his book "The Slipped Disc and the Aching Back of Man" backs up this statement:

"The disc is not essential to one's anatomy. In the theory, evolved by Schmorl in 1928, he suggested that the disc or nucleus pulposus may possibly act as a cushion or buffer between each two opposing vertebral bodies, but I, from the results of my research, have been able to demonstrate that it is the annular ligament which is the real cushion or buffer and not the disc or nucleus pulposus. To help to confirm my theory I was able to demonstrate in the pathology department that many of the specimens who had died of various diseases had no discs at all and yet during their lifetime had had not history of back trouble whatsoever. I thus concluded that the annular ligament can perform its function as spring or buffer in the spinal column quite satisfactorily without the disc. The vertebral bodies do not come together in its absence because of the presence of the annular ligament and that of the articular processes or facets of the opposing vertebrae."

THE STRUCTURE

We have just examined how the vertebral bodies are stacked on top of each other and observed how the disc complex varies when there is no weight bearing (the annular ligament is flush with the sides of bodies), to weight bearing (the annular ligament protrudes over the sides of the bodies), to the spine bending forward (the annular ligament still protrudes but now the bodies are closer together at the front than at the rear and the gap between is wedge shaped resulting in the disc being squeezed backwards under great pressure.)

My two drawings (on page 36 and page 37) show how the vertebrae (bodies and arches) are arranged on top of each other. At the rear of every disc complex and running close by are two nerve roots that leave the spinal cord before supplying the left and right hand sides of the body. Two of these are shown in the drawing. These supply the left side of the body. The two supplying the right side are on the far side of the arch.

Pay particular attention to these nerve roots, the description given to nerves as they leave the spinal cord. They run close to the disc complex that normally is in a state of movement, coping with daily activity. A source of potential danger? Yes!

Examine how the facet joints mesh lightly together. As I have already said their function is to guide the spine when bending and to limit

The Upright Spine

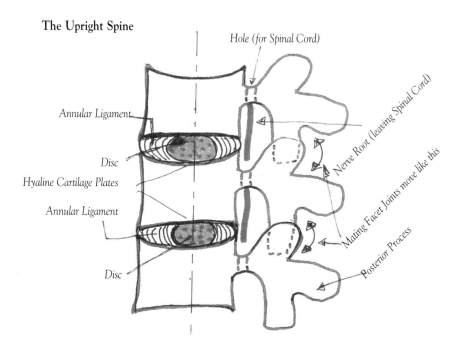

rotation of the spine. Without them the spine would wobble about all over the place.

Now examine what happens when the spine bends forward. (*see opposite*)

With a healthy back the disc complex returns to the position shown in the first drawing, when the spine straightens into the upright position. However, continuous bending forward, as in the second drawing, will in time result in the bands of ligament that make up the annular ligament being over stretched. They lose their elasticity and become thinner due to stretching. They fail to return to their original position and remain permanently protruding backwards, something like the bulge shown at the rear of the top disc complex in the second drawing. This is what's known as a swollen (bulging, protruding) disc. With time further deterioration takes place. The swelling increases, makes contact with a nerve where it leaves the spinal cord and can cause pain.

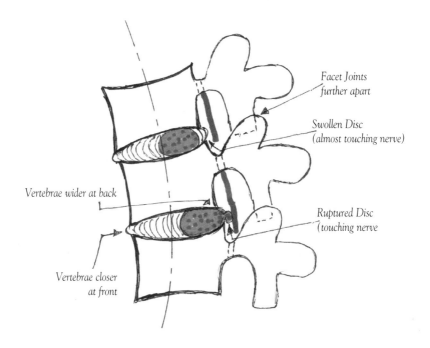

As the rear of the annular ligament continues to be flexed backwards and forwards from an S to a C position the bands of ligament begin to crack. An analogy could be made with bending a piece of metal backwards and forwards on a continuous basis. Eventually due to metal fatigue the metal will become brittle, will fail to return to its exact position after bending and will break.

This is what happens with the annular ligament. The bands of ligament tear and the pressurized jelly-like material of the disc is ejected through the sharp tear. This is known as a slipped (prolapsed, herniated, ruptured) disc and can result in great pain. Notice I use the word "can."

A study carried out in 1990 by Scott Boden of the George Washington University Medical Centre examined pain free people under sixty who never had any back pain or sciatica (leg pain caused by spinal malfunction). Slipped discs are regarded as one of the main causes of back pain. This may be true. However MRI scans given to the 67 pain free people used in the study showed that 20% had slipped discs, 50% had swollen discs. Over sixty years of age the 20% becomes 30% and the 50% becomes 80%. In this age group almost everybody had some age related degeneration of the disc complex. A study in 1994 by Michael Brant-Zawadzki of Hoag Memorial Hospital in California produced similar results.

It is extremely important to bear in mind that if an MRI scan diagnoses that a back sufferer has a slipped disc all it is doing is confirming the disc has slipped. It is by no means confirming this is the cause of the pain. Carrying out invasive surgery based solely on the result of an MRI scan can have disastrous consequences.

Consider the following appalling scenario that is happening on a frequent basis:

A patient, suffering from acute back pain and sciatica, attends an orthopaedic surgeon. A ruptured disc is diagnosed. The patient is sent for an MRI scan which confirms the diagnosis. Based on this scientific evidence, which confirms the "hands on" diagnosis, an operation is carried out to remove the extruded part of the disc pressing on a nerve leaving the spinal canal. The fact that 20% of all pain free people under sixty years of age have ruptured discs is ignored. The operation is a total failure with the pain level even greater than before. Faulty facet joints, as mentioned above, can also cause sciatic pain. A second operation is suggested. The patient, in a state of hopelessness, traumatized and in great pain, will readily agree to anything that shows even a glimmer of hope of alleviating the pain. But will this second operation be any more successful? I have met hundreds of back sufferers who have had multiple unsuccessful operations. One of those I will talk about later had thirteen operations.

Most back suffers depend on others to help them. They attend physiotherapists, chiropractors, osteopaths, acupuncturists, etc., who are usually extremely successful in relieving the symptoms of pain. They offload their chance of pain free living on to somebody

else, with no thought of helping themselves. After a number of sessions they experience great relief and without a thought go back to their normal way of living, driving, working at a desk, sitting, bending, carrying large handbags, lifting heavy loads and abusing their backs on an hourly basis. All the good work is undone and back they go to their chosen expert on back pain relief. The merry-go-round goes, well, round and round. I meet many back sufferers who attend three, four and five courses of treatment yearly. They are treating the symptoms, not the cause.

THE SPINAL CORD

The spinal cord is the elongated nearly cylindrical part of the nervous system that runs from the brain down the back of the spine. The vertebrae form a continuous channel called the spinal canal through which the spinal cord runs. (It could be compared with a multi core telephone cable with the various insulated wires being replaced by nerves.) It runs through the holes in the vertebral arches and at the bottom of each hole a pair of nerves leave the cord to supply the right and left hand side of the body. The function of the cord is to link the brain with the nerves of the body, relaying information to and from the brain.

Three membranes, collectively called the meninges, which are the protective membranes of the brain and the spinal cord, enclose the spinal cord. A watery fluid, the cerebral-spinal fluid that protects the brain and cord from damage by acting as a fluid cushion, also surrounds the cord

You may have experienced pain immediately after sneezing or coughing. This physical act may cause a sudden sharp increase in the cerebral-spinal fluid pressure resulting in a pressure wave running down the spinal cord, causing contact with a disc complex with resultant pain.

It is interesting to note that the spinal cord extends only as far down as the L2, the second lumbar vertebra. From this point individual nerves leave the bottom of the cord and extend down to supply the lower parts of the body. These nerves have a special name, "cauda equina" or "horse's tail".

FACET JOINTS

A large percentage of back pain is due to facet joint malfunction and so it is extremely important you gain an appreciation of how they work.

As can be seen from the drawing the facet joints lock adjacent vertebrae together (Something like the action in children's leggo building blocks.) During bending they make fleeting contact and so guide the spine in the vertical plane. Without them the spine would be unrestricted and could wobble all over the place. They also limit how much the spine can bend.

The facet joints, particularly in the lumbar spine, help limit rotation or twisting of the spine. The annular ligament allows this movement in order for the body to rotate. The facet joints restrict too much rotation, which could cause damage to the annular ligament.

Degeneration of the spine is a natural occurrence and can start at an early age. The moisture content of the disc complex gradually becomes less. This causes the disc complex to shrink, resulting in the gap between the vertebrae decreasing. The facet joints are brought closer together and, rather than making fleeting contact, begin to bear down hard on each other. Friction roughens the surfaces, the capsules (the strong elastic sleeves that bind the joints together, absorb blows passing through the joints and prevent bone-to-bone jarring of the surfaces of the joints) produce excess lubricating fluid which can result in the capsules becoming bloated, compressing the nearby spinal nerve with resultant pain and sciatic pain down the leg.

The facet capsules are equipped with a number of nerves that pick up any malfunction of the facet joint and relay it to the brain. The surrounding muscles then contract to protect the joint. This can cause inflammation of the joint by reducing the flow of blood through the capsule. As the joint becomes more inflamed it sends off more signals to the brain and so the cycle continues.

This is why it is important to be as active as possible throughout a bout of back pain. Signals to the brain are overridden; the contracting muscles become lax and allow the blood to flow normally again.

There is no danger of doing further damage, unlike when taking painkillers. Pain will prevent this happening. Painkillers eliminate pain, give a false sense of wellbeing and allow the back sufferer to continue to abuse the spine and cause even greater damage. Because of this I am completely against pain killers and deem it essential that if they must be taken the back sufferer be confined to bed.

A bloated facet joint capsule pressing on a nerve can cause pain similar to that caused by a ruptured disc. It is one of the many reasons why **"Determining the cause of a given individual's pain often remains more art than science."** (Ref. Scientific American 1998.)

MUSCLES

Muscles are fibrous tissue controlled by the brain with the ability to contract, producing movement or maintaining position. Body movements are generated through the contraction and relaxation of specific muscles.

Some muscles, like those in the back or arms for example, are voluntary muscles controlled by the brain and bring about such movements as bending or raising a hand. These are known as skeleton muscles. Other muscles are involuntary and function without conscious effort. Blinking or the pupil constricting when exposed to bright light are examples. These are known as smooth muscles and derive their name from the fact that, unlike skeleton muscles which have striations (appearance of parallel bands of fibre on the surface), the surfaces are flat.

SPINAL MUSCLES

The muscles of the spine work like puppet strings. They pull on levers that are part of the vertebrae in order to make the spine move. As we have seen already, each vertebra has three levers at the back called processes, (one posterior and two transverse). Without the dynamic support of the muscles the spine would fall over. When it threatens to fall, stumbling for example, the brain is alerted and the muscles are instructed to bring it back to the vertical.

The muscles running down the back of the spine work in conjunction with the abdominal muscles. When the spine bends forward, the two sets of muscles combine to allow a smooth movement. The stomach muscles take the weight of the upper body and progressively lower it into the required position. Simultaneously the back muscles, under tension, slowly pay out the upper body weigh.

When straightening up the process is reversed. The back muscles pull down on the levers (posterior processes) they are attached to and so the spine is pulled upwards. The stomach muscles all the while lengthen and allow this to happen.

Apart from their part in allowing the spine to bend forward, strong stomach muscles are extremely important for combating back pain by involuntarily preventing the seated spine from slumping into a "C" position.

DYNAMIC STABILITY OF THE SPINE

While the various static devices such as custom tailored chairs, car seats and pillows keep the spine in perfect posture during sitting and sleeping, it is also important to train the back to deal with the various dynamic activities which are part of our daily lives. The best way to achieve this is by core exercises.

The body's core is the area around the trunk and pelvis. This is where the centre of gravity is situated. A strong well developed core forms a rigid foundation for the spinal column and provides a stable platform for work activities and sport. With good core stability the muscles in the pelvis, lower back, hips and abdomen work in harmony and provide support for the spine during most activities.

A weak core can make you susceptible to poor posture, low back pain and muscle injuries. Strong core muscles provide the support needed to combat injury and resultant pain.

Core strengthening requires the regular and proper exercising of the body's 29 core muscles. An ideal form of core exercise is provided by the use of an over-sized Swiss ball that requires the use of the core muscles to maintain balance. The use of this type of unstable platform is better than a rigid exercise machine which exercises only one isolated group of muscles at a time

My approach to pain free living is different; it's extremely simple, my own creation, works for me and does not include exercising. I meet many people on a constant basis who proudly tell me that before they leave for work every morning they exercise for a period of time. I say to them "That's great! But then you spend the rest of the day abusing your spine on a constant basis and all your good work is undone." Lack of reactions and guilty looks tell me they agree.

I keep pain free by never, ever, allowing my spine to go from an S to a C position.

I am motivated by the fact that should I not do so, even for a couple of seconds, pain may strike. Of course there are occasions when I fail. I can then spend literally days hoping I have got away with it. (Injuring your back may not result in immediate pain, as I have discovered long ago. It may take hours, days, even weeks before pain occurs. I have met people involved in car crashes who were pain free for months afterwards.

THE CERVICAL SPINE

The vertebrae and the disc complexes in the cervical spine are tiny in comparison with those of the lumbar spine. They are more fragile, seldom at rest, constantly moving up and down, twisting from left to right and so a neck pain is far more difficult to relieve than a lumbar pain. If I were speaking to you now, rather than writing, you would see that my head is never quite still but is moving slightly with every word. The mechanism of the spine is such that pushing the lumbar spine into its correct sitting position results in the cervical spine automatically attaining perfect posture. I will talk about this later when I come to car seats.

Just as bad posture causes pain in the lumbar spine, it also causes pain in the cervical spine. It is extremely important that the head is kept in a position that results in its weight resting on the top of the spine. In this position the disc pressure is the least possible. This is due to a basic principle of mechanics that defines the stress at any point in an object being bent as "force multiplied by distance." When the head is bent forward the stress in the cervical spine is the weight of the head multiplied by the distance from the cervical spine. When the head is resting vertically on the cervical spine distance does not come into the equation and so the stress is zero. I consider this fact to be so important that I would like to give it further consideration.

I am now going to carry out a simple experiment to illustrate that not keeping the head over the spine creates huge stress in the cervical spine. I suggest you do likewise when you have read this.

Hold an A4 sized book in your right hand, assuming you are right handed. Hold it between the thumb and the forefinger, at the bottom right hand corner.

Raise your forearm vertically, still holding the book as illustrated in the first drawing opposite. With the book balanced and the weight of the book going down your arm, it is comparatively easy to hold.

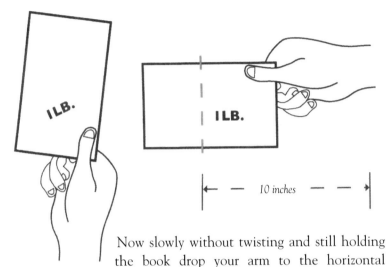

Now slowly without twisting and still holding the book drop your arm to the horizontal position. Even though the book is light, say one pound, many will have difficulty holding it in position. The majority, unable to do so, will not be able to stop the book falling from the horizontal.

The reason for the difficulty in holding the book is due to the fact that, in the first drawing the weight of the book is acting down vertically through the wrist, whereas now, with the arm extended, the weight is still acting down vertically but is now displaced some ten inches from the wrist.

The stress caused by this leverage is calculated by multiplying the weight by the distance from the wrist.

$$1 \text{ lb} \times 10 \text{ inches} = 10 \text{ inch lbs.}$$

Now imagine your head replacing the book and your spine replacing your arm. When your head is right over your spine there is the minimum stress on your neck. When you are sitting with bad posture, working at a desk for example, the centre of gravity of your head is displaced approximately ten inches forward from your spine. The stress in your neck is the same as for the formula for calculating the book:

Weight of head x 10 inches = stress in neck

Have you any idea what your head weighs? It's approximately 14 lbs. Now do the sums:

14 lbs x 10inches = 140 inch lbs.

So the stress in your neck, caused by your head, is fourteen times greater than the stress in your wrist caused by the book and you had great difficulty holding the book. Imagine trying to hold it if the weight was fourteen times greater. You couldn't.

This simple experiment shows why it is extremely important to keep your head vertically over your spine and to always rest your head against the back of the chair when sitting.

Stress

When I use this word I am referring to the force caused by bending a stiff object, a wooden ruler for example and expressed in inch/lbs.

Pressure

When I use this word I am referring to the force in a liquid or gas, a motorcar tyre for example and expressed in lbs/sq. inch.

Medical books use archaic words to describe the various parts of the body, which can be off-putting. I would imagine one of the reasons for this is to avoid the confusion that could occur in using topical words, as these can tend to change their meaning over a period of time.

Some words used in describing the spine and their colloquial meanings are listed below.

Anterior front

Annular ligament elastic compressible closely bonded layers of fibrous tissue that are closely bonded to the vertebrae and enclose the disc

Articular relating to the joints

Cartilage firm flexible connective tissue which covers the ends of joints e.g. chicken bone knuckles. The low friction surface it provides allows bones to move against each other. The surfaces are lubricated by synovial fluid

Cervical kyphosis the arch in the neck

"C" the shape of the spine when the sitting forward with bad posture

Disc (nucleus pulposus) the soft jelly like material lying between every two discs, enclosed by the annular ligament and touching the hyaline cartilage plates embedded in the surfaces of the vertebrae

Disc complex the collective name for the two hyaline cartilage plates, the disc and the annular ligament that separate every two vertebrae

Facet one side of a many sided body

Foramen opening, hole, passage

Hyaline cartilage plates glass-like plates deeply embedded into the lower and upper surfaces of the bodies of the vertebrae. Their function is to provide the discs with nourishment

Inferior lower

Lamina thin plate, scale, layer

Ligament a short band of tough flexible fibrous connective tissue linking bones together

Lordosis the hollow in the lumbar spine

MRI magnetic resonance imaging is a device that uses magnetic fields and pulses of radio energy to provide pictures of the organs and structures of the body

Nucleus pulposus disc

Pedicle root

Posterior back

Process an outgrowth in an organism

"S" the shape of the spine when sitting with perfect posture, which I define as that position of the spine that puts the least pressure in the discs

Sacrum the solid triangular shaped bone on which the spine rests

Slipped disc ruptured, herniated, or prolapsed annular ligament

Synovial fluid lubricating fluid for joints

Swollen (bulging disc) condition of annular ligament when permanently protruding backwards

Thoracic kyphosis the slight forward curve in the thoracic spine

CHAPTER FOUR

SWEDISH RESEARCH INTO DISC PRESSURE

"Poor posture will cause back pain"

ENGINEERING STRUCTURE

The spinal column is like an ingenious piece of mechanical engineering design, controlled by a computer so sophisticated that it may be centuries before man is capable of fully understanding how it works. Yet, the more I researched the cause of back pain, the more I realized the design of the spine had a glaring flaw. It appeared as if it was designed to cope with a way of life that existed many thousands of years ago, when man never sat.

My research told me eighty percent of the world's population suffers from back pain at some time in their lives. (Ref. The World Health Organisation) and as everybody knows, the main cause is sitting with bad posture. I then found it extraordinary to learn that amongst primitive people the incidence of back pain is nearly non-existent. I discovered the reason for this is because they don't sit like we do. They squat, like we did in the distant past. We now sit and suffer back pain.

BACKPAIN AND PRIMITIVE TRIBES

"Backpain is an affliction which affects about 80% of the population in Western countries and sadly the indications are that it would seem to be on the increase. Why is this? With advanced technology and improvements in working methods surely we should expect to have fewer physical complaints. This is not so.

The human body was built for movement, for physical activity, with speed and precision being the important factors. Children born to-day are the same and have the same physical equipment, as did our pre-historic ancestors. Our ancestors chased after animals for meat, climbed trees for fruit, walked miles in search of grain, in other words they were extremely active. Later when they domesticated the wild animals and cultivated fruit and grain they were still very mobile because all work was done by hand. Squatting was the position of choice not only for rest but also for tool making - our first industry.

Thus facility for movement was utilised to the full. Mobility in all joints was maintained and the stronger muscles were used in the work for which they were intended, that is to propel the body in any direction and lift whatever weight was required to be carried.

When our development is traced through the ages, it is noticed that we have carefully and methodically engineered our physical ruin. Technology has been to our benefit but the changes brought about have created problems. We have failed to adapt adequately to those changes through a diminished range of mobility.

Although we are without any documentation about the physical well being of our early ancestors, we can learn much from studying primitive tribes in today's world. For instance the Aboriginals in Australia suffer very little from back pain and it is noted that the low back curve in this tribe is less marked than in Europeans. African tribes also reveal a very low incidence of slipped disk and are seen to have an ample range of movement in the joints of the low back."
(Ref. Back Care by Department of Health)

REDESIGN

If the spine were an engineering structure designed by man many years ago and was now out of date and not able to adequately fulfill its purpose it could be redesigned to take into account any new circumstances. But man, who is now capable of pushing back the boundaries of medical science so far that even a few short years ago they would have been considered impossible, did not design it. Cloning, organ transplants, are achievable on a daily basis. But the redesigning of the spinal column? No!

So we are stuck with it and will have to consider ways of adapting it to present conditions.

BALANCING SYSTEM

These were some of my considerations as I approached the problem of the relief of back pain. I knew that everybody stands and walks with reasonably good posture. If they didn't, if their posture was atrocious they could fall flat on their faces, or flat on their backs. The design of the spinal column is so sophisticated that when you are walking on the footpath and you stub your toe in most cases you don't fall. The reason being the instant communication between the brain and the balancing system in the ears that instantly instructs the large muscles in the back to pull the falling spine back to the vertical. This ingenious system goes into repose when the person is seated and the spine collapses into a forward "C" position, creating great stress in the cervical and particularly the lumbar spine and eventually resulting in pain.

> **"Even though bad posture may not cause any discomfort, continual poor posture will in the long term cause back pain." (Ref. Back Care by The Health Education Bureau)**

I knew that when I sat with bad posture I would experience pain. My pain was the result of a schoolboy rugby injury aggravated by years of sitting with bad posture. Also, I was aware that eighty percent of the world's population suffers back pain at some time in their lives, a universal figure throughout the developed world and that among primitive people the incidence is low because they don't sit, they squat. This knowledge gave me the key to back pain relief.

The most obvious solution was never to sit. This is totally unrealistic and I dismissed it immediately. Sitting is essential to cope with modern living. Ever drive a car, work eight hours at a computer, standing or squatting? So people must sit. Ideally, when sitting, a system must be devised to enable them to do so with the pressure in their discs reduced to the same level as achieved by squatting.

I designed and patented such a system, Spinal System-S, which gives the information to manufacture seating that results in the pressure in the discs in the seated spine being similar to squatting.

I use the word "similar" as I don't know what pressure exists in the discs in the squatting spine. I do know however that the result of sitting with S and squatting is pain relief.

Sitting with S reduces the pressure in the C sitting spine by fifty percent. The following is the result of a Swedish study:

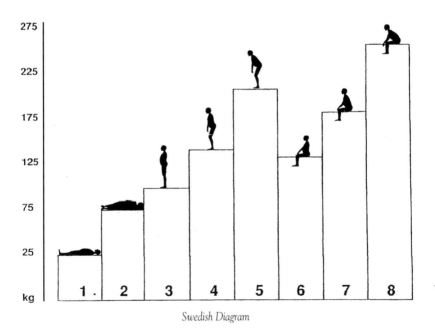

Swedish Diagram

The diagram shows how the pressure, in kilograms per square centimetre, in a third lumbar disc varies in different body positions. This was achieved by inserting a hypodermic needle into the volunteer's third lumbar disc and observing the pressure in the disc in various positions.

As can be plainly seen the pressure is least while lying on your back (no1) and greatest while sitting with bad posture (no.8)

The most interesting fact is that sitting with bad posture (no.8) results in twice the pressure in the disc when sitting with perfect posture (no.6), 250 kg per square inch to 125 kg per square inch. It is also much greater than partially bending forward when standing (no.5).

In my opinion this is an extremely important piece of research work and should figure on every school curriculum. It graphically illustrates why sitting with bad posture should be avoided, a point made to you, me and everybody during schooldays, but never heeded. I think a sight of the above diagram might have achieved more success.

CHAPTER FIVE

DEVELOPMENT OF SPINAL SYSTEM-S

"The distorted part of the spine fits neatly into the void"

Soon after I started making custom designed car seat moulds that guaranteed perfect posture, I was approached by the Eastern Health Board to make wheelchair seat moulds. I call them moulds because they mould individual spines into their own unique S shape, which as you know I define as perfect posture.

SPINAL SYSTEM-S

At this stage in its development my system for measuring spinal curvature, I call it **Spinal System-S**, was still in a crude state. I was still using the same primitive measuring device I designed after leaving hospital. It bore little resemblance to the final Spinal System-S measuring system on which I hold a patent and which is shown on the cover of this book. However I used the same techniques then as I do now for measuring S. I gently push the spine into S and after twenty-five years of measuring the spinal curvature of some ten thousand back sufferers I know with 90% / 95% certainty the S for any given individual. In consultation with the person being measured I then arrive at the exact S. I also determine the depth, width and height of seat and the arm height. Arm height is important as sitting with both elbows correctly supported transfers a large amount of bodyweight from the spine and down the arms, thereby greatly reducing the pressure in the discs. I use all this information to manufacture office, home, or car seats that guarantee **S.**

One of the first people I measured was one of the best-known medical people working in the area of back pain relief and a consultant in one of the main Dublin hospitals. I had been making individually designed back braces made out of aluminium and covered in leather for his patients. I was having some difficulty measuring him as he was fidgeting and talking about lordosis, flexion and kyphosis of his spine. As I was concentrating on what I was doing he realized he wasn't getting his point across. He stood up, dropped his trousers to his ankles, took off his shirt, sat down and continued talking about the three curves of the spine. Seeing my look of amazement, he grinned. "Get used to a naked body. You may have to ask everybody to do this." He was wrong. His striptease was a once off.

Sometime after this I was approached by a middle-aged man who asked me if he could speak to me in private for a couple of minutes. I know it was winter as I remember he was wearing a long overcoat with the collar turned up and a dark wide brimmed hat. He looked vaguely familiar and while I couldn't quite place him I knew I had met him before.

I told him we could speak here as there were no customers present. He asked if we could go into the small room I used for measuring spinal curvature, so I knew then he had been a customer. When we went into the room he asked to see my hands. Embarrassed, I tried to talk my way out of complying. He insisted, so I held my hands out, palms upward. He took off his overcoat, revealing a white clerical collar and in Latin proceeded to bless each hand in turn. He then put on his overcoat and hat, said goodbye and left. During the whole episode I never said a word and apart from speaking in Latin neither did he. I am not a religious person, but the fact that a minister of religion thought so much of my work that he bestowed on me what he considered the great honour of blessing my hands meant this was a definite signpost on my journey to my final destination, the perfection of Spinal System-S.

Another signpost came in the guise of a woman in her sixties, resplendent in a fur coat and gold jewellery. She walked in the door with a pronounced limp and handed me a piece of headed notepaper with the name of a well-known consultant in Baggot Street Hospital (now closed). The notepaper had a crude drawing of a foot indicating it needed an arch support. I told her I knew nothing about feet. She spoke for the first time saying, "My consultant said that if Colm Campbell can't fix my foot nobody can."

At that time a material called thermo-plastic had come on the market and companies were seeking applications for its use. It was supplied in rigid sheets of various thickness. One of the companies had given me several sheets measuring twelve inches by twelve inches by one eight inch thick. The application of heat, warm water would do, caused it to lose its rigidity and become soft and pliable. When it cooled it reverted to its original rigid form. I had been looking at using it for sports injuries, particularly for neck injuries. My thinking was to supply the thermo-plastic in packages. When an injury occurred a battery enclosed in the package could be activated, the heat generated would immediately soften the thermo-plastic and the package could be moulded around the neck, As the thermo-plastic returns to full rigidity in a short time when it is removed from heat, it would then be simply a matter of holding it in position to allow this to happen. The injured person could then be removed to hospital without fear of damage to the spinal cord.

I immediately saw a possible application for supporting the woman's foot. I asked her to be seated and to remove her right shoe. I took the shoe and did some quick measurements of width and length. Using a Stanley knife I cut a piece of the thermo-plastic to the required dimension. I put the rigid piece into a kettle of warm water for less than a minute. When I took the plastic out it had gone as soft as a handkerchief. I placed a piece of two-inch polyurethane foam on the ground, with the thermo-plastic, now flattened, on the top. I told her to press her foot down on the thermo-plastic and held it in position for a couple of minutes until the thermo-plastic had returned to its rigid state. After numerous adjustments I arrived at a perfect arch support for her foot. I fitted the support inside the shoe and she left, without a trace of a limp.

Making wheelchair seat moulds led me into the area of the physically disabled where I worked exclusively for a number of years with various health boards throughout the country. It was very satisfying work and required a certain amount of innovation to cope with some extremely distorted spines.

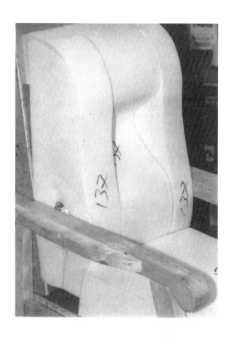

The photograph shows, in the raw state, an example of a chair I made for a person with spina bifida. As you can see the distorted part of the spine fits neatly into the void in the back of the chair. All the weight of the upper body is taken by the back of the ribcage, with no pressure on the part of the spine that fits into the back of the chair. When the chair is upholstered, the space for the distorted spine is barely noticeable.

I expanded into the manufacture of custom designed chairs when the Eastern Health Board asked me to make one for a 42-year-old mentally handicapped quadriplegic with a badly distorted spine, who had been lying on the floor all his life. I went to see him with an occupational therapist from the Health Board and found him abandoned lying in the foetal position on a rug in the corner of a room. Obviously somebody must have been looking after him but every time we called there was no minder present. The occupational therapist had a front door key. Because he was completely incapacitated he was difficult to deal with. Two of us were needed to keep him sitting upright on my measuring system and I also had to measure his spinal curvature. It took some time and several return visits to the Dublin suburb of Chapelizod where he lived, to make the chair. The occupational therapist and myself spent some time getting him correctly seated in the chair. He was now sitting upright for the first time in his 42 years existence. Leaving the room some time later I took one last glance over my shoulder. His right hand resting on the armrest caught my attention. Unable to speak, paralyzed, but to this day I'm certain I saw a slight quiver in his thumb, indicating the chair was successful.

The success of this first chair resulted in more orders that greatly improved the quality of life for many physically disabled people. This particularly applied to those who spend all their waking hours confined to wheelchairs. A wheelchair is designed as a vehicle to get a handicapped person from A to B. The fact that it can be folded to allow it to be stowed, in a car boot for example, dictates that support for the spine is virtually non-existent, resulting in the person slumped, totally immobile, for anything up to twelve hours at a time. Imagine the effect this would have on a perfectly healthy spine which can become uneasy even after a relatively short period of sitting without being able to make minor adjustments to the sitting position. Imagine also the huge benefit, mentally and physically, to a wheelchair bound person who can now sit in perfect posture in an electrically operated reclining chair and can independently control the position by means of a finger controlled zapper.

Simultaneously to this work and in conjunction with the Central Remedial Clinic (a medical clinic specializing in cerebral palsy) I was involved in a programme of designing custom made seats for insertion into buggies for infants suffering from cerebral palsy. This led to me forming a joint company with the Clinic.

The obvious next step in the development of Spinal System-S was to expand into manufacturing office and lounge chairs for the general public.

The following are some of the letters I received from grateful customers:

"Just a few lines to thank you for my chair and to let you know that it really works. I honestly never thought anything could help ease the pain in my back, as it was so badly curved inwards and weak from muscular dystrophy.

Anyway it is marvellous and I am also thrilled that the chair looks so nice. It is important to me that it doesn't look different from what an ordinary person would use. Yet it does it's own special job as well.

Once again I thank you for my chair and all the trouble you took to give me exactly what I asked for."

"Just to let you know we are delighted with Jason's chair. There hasn't been a mark on his spine since he got it and he really enjoys the way it reclines. It seems to be doing the job that we needed. We should have bought it long ago. Thank you for all the adjustments you had to make to the chair to accommodate him. We know it was a little difficult. Jason is very comfortable looking in the chair and seems very happy while he is in it. We would certainly recommend your type of chair to anyone with back problems as it definitely doing good for Jason." (Jason is a mentally handicapped quadriplegic.)

"To tell you as promised that the chair is wonderfully comfortable. I can sit for several hours in it, so easily rising from it and not at all crippled with the spinal pain when I do get up from it.

It is fantastic the difference it makes supporting as it does the whole spinal column. The soft half seat for the stiff hip side means that even the first steps I take on rising are without hesitation and pain.

God bless your hands and the intelligence which inspired you in designing the chair."

CHAPTER SIX

INVASIVE SURGERY

"No matter how successful the operation is your back will never be normal again"

This research work during the early eighties gave me an appreciation of the anatomy of the spinal column, how the pressure in the discs varied hugely from one body position to another and the potential danger that exists with the nerves close to the disc complex bulging in and out due to various activities on one side and the facet joints moving up and down mainly in the vertical on the other.

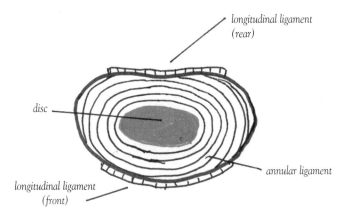

longitudinal ligament
(rear)

disc

annular ligament

longitudinal ligament
(front)

When I stood up to answer that telephone call some twenty-five years ago my fifth lumbar disc prolapsed. Having read the foregoing you will now realize that the soft jelly-like material that when young has a ninety percent water content had burst through a tear in the annular ligament, struck the sciatic nerve and I hit the floor unconscious. The disc had ruptured (herniated, slipped, or prolapsed) and like toothpaste squeezed out of a tube cannot be put back inside the annular ligament. You now realize this is what most people refer to as a "slipped disc," and are under the impression that the disc can be "put back." You know it cannot.

Once the disc remains in contact with the polished cartilage plates, on the surfaces of the vertebrae, it receives nourishment and remains healthy. When contact is broken it dies.

This last statement holds the key to the elimination of a great deal of the world's back pain. I will talk in detail later about Alexander Walker-Naddell, an orthopaedic surgeon and a fellow of the Royal College of Surgeons, who perfected a non-invasive technique for breaking the contact between an extruded disc and the cartilage plates providing it with sustenance. In spite of the fact he was hugely successful and patients came to him from all over the world to avail of his non-surgical removal of a slipped disc, fellow members of his medical profession shunned him. This was because he didn't peer review his research.

I now want to bring you through the whole shadowy area of invasive surgery, giving my own observations as a layman and of far more importance the findings of eminent medical experts working in this field. You will find them interesting.

The recognized medical procedure for the removal of a prolapsed disc is an invasive operation. I have spent the last twenty-five years working exclusively in the area of back pain relief. I have measured the "S" shape of something in the region of ten thousand spines. On top of this I would have spoken to many more tens of thousands of back sufferers about the history of their back pain. Out of this huge number, in the region of forty thousand, the percentage of successful operations I have encountered is less than one percent. This is not a scientific study insofar as most of the back sufferers came to me rather than me carrying out a recognized sampling procedure whereby people who had back operations would be approached to determine success rates. Nevertheless it is quite alarming and if nothing else will encourage back sufferers to stop handing over the solution to their back pain to a third party, take responsibility for their own problem, concentrate on non invasive treatment and forget about the quick fix.

A large number of back sufferers have informed me that before their operation they were told the success rate was 80%. The extraordinary thing is most of them hold no bitterness towards the surgeon for quoting this huge success rate. I would guess that ignorance is the main factor for this benign attitude. They never, ever, ask the surgeon to define what is meant by the word "success." They will grasp at any hope of becoming pain free, concentrating exclusively on the positive, all logic temporarily abandoned.

If they had been informed they would know that surgery is necessary in only a small percentage of cases.

There is the huge possibility they would have refused an operation like I did and would now be pain free.

Professor Deyo, Department of Medicine, University of Washington states:

"Studies using repeated MRI revealed that the herniated part of the disc often shrinks naturally over time and about 90 percent of patients will experience gradual improvement over a period of six weeks. Thus only about 10 percent of patients with a symptomatic disk herniation would appear to require surgery. And because most pain is not caused by herniated disks, the actual proportion of all back pain patients who are surgical candidates is only about 2 percent."

Given that only 2 percent of back sufferers qualify for surgery, would you agree to go under the knife based on these odds?

"Before you decide on back surgery consider getting a second opinion." (Ref. The Mayo Clinic, on their website) Go to my website www.back-shop.com and you will find a link to their back pain pages.

If you were looking for, say, a contractor to carry out work on your house a normal procedure is to ask for three quotations. Yet for something more important, dangerous reconstruction work on your spine, you accept the first contractor. My advice is to go one better and, if possible, ask for four. On a number of occasions I have been approached by frightened, terrified would be a more appropriate word, back sufferers who told me they had been advised that unless they had an operation they would be paralyzed for life. I advised a second opinion. On every occasion they came back to thank me for my advice. The only one I was able to keep track on is the wife of an acquaintance of mine who approached me about his wife's neck pain. That was five years ago. She didn't have an operation and is now doing fine.

The Mayo Clinic also states:

"Long term outcomes also are often similar following less invasive treatments."

This statement by the renowned medical clinic in the USA. is a huge step forward, not for mankind, but for "medicinemankind."

However I disagree with it. In my opinion, based on my own personal experience, my research and the knowledge gained over the last twenty-five years working in the area of back pain relief, alternative treatment is infinitely more beneficial than surgery. First of all it does no harm. It is successful when entered into in whole hearted manner (I will talk about this later), is non-invasive and the horrendous consequences of botched operations are avoided.

Hamilton Hall MD FRCS(C), an internationally recognized authority on the treatment of back pain and a member of the International Society for the Study of the Lumbar Spine, in his book "The Back Doctor" states:

"No matter how successful the operation is, your back will never be normal again; surgery creates scar tissue, which doesn't exist in a normal back."

Dense scar tissue forms around the nerve leaving the spinal canal, irritating the nerve, which can result in pain, pins and needles and numbness. This "surgical garbage" even has a name, "archnoiditis."

Hamilton Hall also contends that:

"If you are to undergo surgery, you must be prepared to make permanent changes in your lifestyle after the operation, doing the exercises and adopting the daily postural habits that will maintain your back in good condition, free of pain."

I will talk later about the fact that most back sufferers hand over the solution to their problem to a third party, be it a surgeon, physiotherapist, chiropractor, osteopath, etc. I have met thousands of people over the last twenty years who have had non-surgical treatment for their backs and after treatment by chiropractors, physiotherapists, osteopaths, etc., their back pain has been greatly relieved. But these experts are only treating the symptoms. The permanent solution to the cause of your problem lies in your own hands, in your daily postural habits, as Hamilton Hall is indicating.

It is my contention that "doing the exercises and adopting the daily postural habits that will maintain your back in good condition" will in the vast majority of cases eliminate any need for even considering the need for invasive operations.

"In any case, spinal surgery can't really cure your back-at least not in the sense that an appendectomy can cure appendicitis. Alterations that take place in your spine during surgery may even cause or contribute to new back trouble at some time in the future." (Ref. The Back Doctor by Hamilton Hall M.D.)

In writing this book I am making three points. The first one is:

You cannot hand over the solution to your problem to a third party. You can seek help from a third party. But the permanent solution to relieving back pain lies in your own hands.

I quote a passage from "The Slipped Disc and the Aching Back of Man" by Alexander Walker-Naddell FRCS:

"**For all these reasons, there was growing concern in the neurosurgical department that the results from these operations did not warrant the inherent dangers of the operation. The senior surgeon, Mr. Eric Paterson, decided that we should recall 50 patients, who had been operated upon in our department, disregarding totally the surgeon involved and that, of course, would include some of my own patients. I was asked to assess the condition of each patient. At the end of five months, I submitted a detailed report of my findings and after careful examination in the department we discovered out of the 50 patients only three could be considered to show good results.**"

Also

"**So we agreed to bring in another 50 patients who had had laminectomy in our department. Once again there was overall a poor result-this time only two patients could be considered satisfactory: even at that, we were assessing for an acceptable day-to-day lifestyle.**"

This represents 4% of patients who now had "an acceptable day-to-day lifestyle." If 4% of operations are reasonably successful the corollary is that 96% are unsuccessful. And yet some surgeons claim an 80% success rate. Of course all of these percentages are dependant on your definition of the word "success." How is "an acceptable day-do-day lifestyle" defined? How do some surgeons define the quality of life in referring to an 80% success rate? And so on. My definition of a successful operation is:

One that permanently restores the body to the same healthy condition that existed before the malfunction occurred.

Rather than give you my layman's definition of "laminectomy," mentioned in the second quotation above, I will quote Walker-Naddell FRCS who is referring to the surgical operation on a slipped disc in his book, "The Slipped Disc and the Aching Back of Man,"

(Lamina is the name given to the area of bone that surrounds the back of the hole in the arch.)

"As the name suggests this operation involves cutting through the lamina of the appropriate vertebra in order to gain access to the offending disc. The surgeon inserts a thin, long -shafted instrument with a small spoon like head, through the tear in the annular ligament, which is lying between two opposing vertebral bodies. He then gouges out the disc or nucleus pulposus from the ligament. The attendant risks of this operation are obvious where the surgeon is operating so close to the spinal cord. He is also working "blind," as it were and can never be 100% sure of clearing out every particle of disc. Most importantly, the fact that a lamina has been cut through undoubtedly causes a permanent weakening of the spinal column."

A long-term follow-up study of 280 patients, performed by Henrik Weber of Ullevaal Hospital in Oslo and published in 1983 raises serious questions about the enthusiasm for surgical intervention."

Spontaneous disc complex prolapse doesn't happen. Laboratory tests show that when a spine is subjected to progressive loading the bones eventually crumble but the disc complex remains perfect. Undoubtedly my rugby injury, followed by years of abusing my spine, inevitably culminated in the disc breaking through the layers of the annular ligament. My injury resulted in huge pressure being generated in the disc, to such an extent it probably badly strained the first layer or two of the concentric layers of ligament tightly encasing it. The annular ligament then had a permanent weakness. The number of layers enclosing the disc had effectively been reduced. I continued to play contact sport, continued to abuse my spine on a daily basis and so slowly over the years one after another of the layers of ligament was penetrated by the disc constantly subjected to bouts of high pressure. Eventually the disc meandered almost through all the layers and the simple action of standing upright from a seated position generated sufficient pressure in the disc to enable it to burst free.

Most back sufferers I meet who have had operations for "slipped discs" tell me that the pain was so great they would have done anything to relieve it, that they were told an operation was the only option and the horrific consequences of not having one. I refused an operation because I was too scared to have one. I am eternally grateful that my decision was motivated by fear, after seeing the result of my bedside neighbour going under the knife. Had I been given a doom-laden scenario, resulting in a state of painful hopelessness, knowing nothing about back pain and thinking my life was at an end, I would have meekly submitted to almost anything and would in all likelihood be still in pain.

Nobody told me that:

"Recent studies show that even for patients with a herniated disc, spontaneous recovery is the rule." (Ref. Professor Richard Deyo, Department of Medicine, University of Washington) So the fact that I recovered and am now completely pain free, is not unique. It is the rule. During my stay in hospital nobody mentioned this "rule."

He also says that most people with acute back pain simply get better. He backs up the Mayo Clinic's statement about the success of non-invasive surgery with this statement:

"Most patients with acute lower back pain simply get better - and quickly. A study comparing treatment outcomes found no differences in functional recovery times among patients who saw chiropractors, family doctors, or orthopaedic surgeons."

This finding, in conjunction with a similar one by The Mayo Clinic, begs the question of why people readily agree to surgery.

Another extremely important point to be borne in mind is that determining the cause of back pain can be extremely difficult. Surgery based on a wrong diagnosis can be disastrous.

"Simple muscle soreness from physical activity very likely causes some back pain, as does simple wear and tear on discs and ligaments that creates micro traumas to those structures, especially with age. Determining the cause of a given individual's pain,

however, often remains more art than science. With spontaneous recovery the rule - once serious disease is eliminated as a factor - pinpointing an exact cause may not be even necessary in most cases." (Scientific American 1998)

I remember attending a back pain seminar in University College Dublin some years ago. One of the guest speakers was a doctor attached to a well-known insurance company in the UK. He spoke about a phenomenon that often occurs when companies are going into liquation - sudden outbreaks of back injuries among the workers. Most of these are malingerers, seeking extra redundancy money. However in every case a deal had to be done by the insurance company. When somebody claims to be suffering from back pain it is impossible to prove otherwise. My simple philosophy is that if it can't be proved that a person is not suffering back pain how can it be determined for certain what is causing the pain.

Another lecture at the seminar was given by a Dr. Johnson, an engineer still in his twenties and working full time on research on the lumbar spine. It reminded me of my crude experiment using wooden blocks. Somebody in the audience asked a question about the cervical spine. He stood motionless, as if transfixed. The seconds passed. I'm certain they turned into minutes. Then, "I'm sorry. I haven't got that far yet." There wasn't even a suspicion of a titter, a measure of how the audience viewed his honesty.

The sudden outbreak of excruciating back pain, even among those who have been battling bouts on a regular basis, can be terrifying. Many of you reading this book will concur. Your immediate, reflex action is to get a medical opinion. What you are told during this consultation can often scare the daylights out of you. A back sufferer came to me recently, a woman in her forties. I knew she was frightened, asked her what was wrong with her back. She gave an Oscar winning performance of a silent movie actress showing fear. Two eyes darting right, two left. Bending conspiratorially she whispered, "I've cervical spondylosis." She didn't know that "cervical" simply means "neck." "spondyl" means the spine. "osis" means "disease of" or more loosely "something wrong with." To a layperson many medical terms are gobbledegook and can be quite frightening especially to someone in pain and extremely worried.

Hamilton Hall FRCS in his book "The Back Doctor" talks about

"In my private life I speak English, just as most of my patients do. In my professional life, however I communicate with my colleagues in medical jargon- a language I call Doctor. Like the jargon of other occupations, Doctor has its place. But it doesn't belong in a doctor-patient consultation."

He goes on: "The game of Speaking Doctor is especially unfortunate when it's played with back patients, because their treatment depends for its success on their thorough understanding of what's causing their pain, what they can do about it and why they needn't be afraid of it. If your doctor uses a term you don't understand, ask for a translation."

Why patients are fobbed off with this type of arrogance never ceases to anger me. It creates fear and anxiety in the average person who knows almost nothing about the spine. This results in depression and cultivates such a feeling of hopelessness that the patient will agree to any treatment suggested by the GP. No thought is given to the consequences. And bear in mind I have already quoted an article from Scientific American that states, "Determining the cause of a given individual's pain, however, often remains more art than science." So how could a GP without any diagnostic equipment possibly give a diagnosis?

It is my experience that the three most common reasons for back pain given to patients are "wear and tear," "degeneration of the spine" and a "touch of arthritis."

Every person suffers from "wear and tear" and "degeneration of the spine." They mean the same thing and are another way of saying you are getting older. There are many visual signs of aging: hair going grey, wrinkles, stiffness in body joints etc. In relation to the spine they refer to the fact that the moisture content of the disc complex decreases with age. As you already know this results in the distance between the vertebrae decreasing, resulting in the facet joint surfaces starting to grind causing stiffness and pain: the proverbial "touch of arthritis."

I have already mentioned an extremely interesting 1990 study by Scott Boden of the George Washington University Medical Centre that looked at people who never had back pain or sciatica, (leg pain from low back conditions). The study showed the fallacy of diagnosing back pain based solely on an MRI scan that shows the person has a ruptured disc. Ruptured discs are often blamed for back pain. But the study showed the following astonishing results:

> **20% of pain free people under 60 have ruptured discs.**
> **50% of pain free people under 60 have bulging discs.**
> **and**
> **30% of pain free people over 60 have ruptured discs.**
> **80% of pain free people over 60 have bulging discs.**

The fact that a person has back pain and an MRI scan shows that a bulging disc or a ruptured disc exists only proves that that person has a bulging disc or a ruptured disc. It by no means proves that the disc is the cause of the pain.

"Abnormal MRI scan from pain free subject illustrates one of the great pitfalls in diagnosing a cause for low back pain. Numerous studies have shown that large numbers of asymptomatic people (pain free) also have disc bulges or herniated discs." (Scientific American 1998)

All the foregoing is beautifully summed up by Sarah Keys (Ref. Back Sufferers' Bible, which is easily the best book I've read on the subject and would strongly advise you to read) when she says,

"The more selective surgeons have strict guidelines and operate only if there is evidence of the nerves in the saddle area and legs not working properly. Pain alone is no reason for opening a back and removing a disc. It is too emotive and too selective. And besides, there are many other spinal disorders which can produce

similar pain. It is too awful if a disc is removed and the pain is still there - and it happens all the time."

Speaking about Disc Surgery she writes:

"The fact that the whole metabolic climate inside an inflamed segment contributes to the irritation of a nerve root may explain why removing a disc with surgery is so often unsuccessful. Some figures estimate that 50 percent of operations for a slipped disc leave the patient worse or at least no better. Removing a disc may not be removing the problem, it may be worsening it. If indeed the facet is the main source of pain, wholesale disc removal obliterates the disc space and brings more pressure to bear on the facets. After the operation the leg pain is much worse - which is very depressing after all you have been through. No sooner are you upright than all your symptoms return, as bad as they ever were. Sometimes you hear of repeat surgery two or three weeks later to operate on another level."

Surgical operations are carried out on every part of the body. Afterwards many patients are on medication for life. This does not happen with back operations, normally the only after care may be a course of physiotherapy. Then, that's it, off you go, we've fixed the problem. No thought is given to what caused the problem in the first place, that it hasn't been eliminated, will return again to gnaw away at the weakened spine and so the cycle is repeated.

I have met hundreds of back sufferers who have had operations in many different countries throughout the world. The same almost complete lack of advice on aftercare and what may be the cause of back pain and its elimination is universal.

The worst case of repeated operations I have come across was a woman, thirty years of age, on two crutches, who came to me by ambulance. She had thirteen operations inside three years. Later, after I had made her a reclining chair for home, I began to think this couldn't be so. Maybe it was three in thirteen years. But no, she returned about a year later for an office chair and I was able to verify that it was thirteen operations.

I am against elective back surgery, except for some critical cases, for example, where nerves in the groin and legs, are not working properly. However if strict instructions were given after surgery on keeping the now flawed spine in its correct "S" position at all times I'm certain my experience of less than 1% successful operations could be improved upon.

The only details of research I know into the success rate of back surgery were the two assessments, just mentioned, carried out by Walker-Naddell into two groups of patients with fifty in each group. The finding in one group was "three (6%) could be considered to show good results" and from the second group "two (4%) could be considered satisfactory; even at that, we were assessing for an acceptable day-to-day lifestyle." I interpret this as zero success rate in each group, applying my definition of a successful operation as "one that permanently restores the body to the same healthy condition that existed before the malfunction occurred."

Even applying Walker-Naddell's loose definition of success his findings of 6% in one group means 94% were failures and 4% in the other group means 96% were failures.

It is my absolute conviction that people contemplating spinal surgery should be made aware of the precise risks involved and so be in position to make informed decisions. To achieve this a scientific study, based on agreed criteria, should be carried out by the medical profession which would evaluate operations and grade the various "success" rates over a period of time. At the moment everything is hearsay.

I have concentrated briefly on the mechanical details of the spinal column in order to give you an appreciation of your back and how abusing it on a daily basis will eventually result in pain.

The most common sites of a slipped disc lesion are in the lumbar spine:

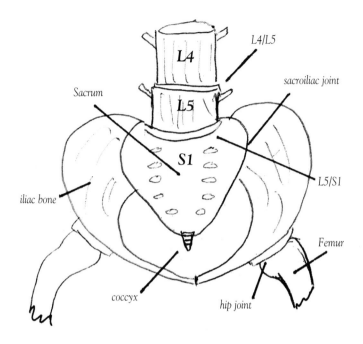

L5/S1 is the most common. This is the lowest disc and so the intervertebral disc pressure is greatest, also the sharp angle between L5 and S1 (See following diagram.) results in huge stress in the rear of the annular ligament which at this point is not protected by the rear longitudinal ligament. This area is closest to the spinal nerve roots and so a prolapsed disc will strike the sciatic nerve and the pain can radiate through the buttock and down the leg. (L5 is the lowest lumbar disc. S1 is the sacrum the triangular bone on which the spinal column rests.

L.4/5 is the next most common. The prolapsed disc strikes a nerve that causes all the muscles supplied by it to go into severe spasm, resulting in severe pain in the mid back. These muscles can in extreme cases pull the spinal column towards them. This sideways bending of the spine is called lumbar scoliosis and can result in an apparent shortening of the leg on the side opposite the spasm.

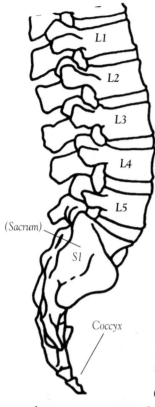

L.3/4, L.2/3 and L.1/2 are not common and are usually the result of an accident

A slipped (ruptured, prolapsed, or herniated) disc in the cervical spine is not as common as in the lumbar spine and is usually caused by an injury such as a "whiplash" causing the head to violently jerk backwards and forwards. This can tear the annular ligament resulting in the disc (nucleus pulposus) oozing out through the tear and coming in contact with a nerve root with resultant pain. The site of the pain will depend on where the annular ligament has been torn and which nerve in the cervical column has been struck.

C.6/7 is the most common and results in acute pain at the base of the neck, which can radiate down the arm and cause "pins and needles" in the fingers.

C.5/6, C4/5, C.3/4, C.3/2 prolapsed discs result in acute pain in the affected areas. The muscles go into spasm pulling the neck towards the affected side.

Slipped disc lesions rarely occur between C1 and the base of the skull.

Walker-Naddell FRS (In his book "The Slipped Disc and the Aching Back of Man") says:

"From a clinical point of view a slipped disc condition in the cervical region should be divided into three specific areas:

1 Disc lesion between C.6/7

2 Disc lesion between C.2/3, C3/4, C5/6

3 Disc lesion between the base of the skull and C1

However there may be an overlapping of clinical features in each, which may lead to early confusion in coming to a firm diagnosis. The two most common sites for spinal disc lesions in the neck are between C.6/7 and between the base of the skull and C1

1 Symptoms of Disc Lesion between C.6/7
The prolapsed disc in this position leads to very acute pain at the base of the neck on one side or the other for it is unusual for nerve roots on both sides to be struck at the same time. The muscles supplied by the affected nerve go into spasm which may extend partially up the neck and, at times, downwards affecting the muscles in the area of the scapula (One of two large triangular shaped bones that form the back of the shoulders.) The nerves supplying all the muscle power to the arms can be struck resulting in marked loss of power in the hand as a whole and " pins and needles" affecting every finger of the hand.

2 Symptoms of Disc Lesion between C. 2/3, C3/4, C4/5 and C.5/6
The prolapsed disc, in any of these sites striking the nerve roots, leads to acute pain in the affected area with marked spasm of muscles supplies by these nerves. The affected muscles in spasm will hold the neck in a fixed position pulling it towards the affected area. This is medically called "scoliosis" (When it happens in the neck it has its own name and is referred to as "torticollis") Pain may extend downwards to the base of the neck and upwards to the base of the skull. Unfortunately, quite often there are no noticeably particular features and often a spinal lesion here is not diagnosed, though in my opinion, most cases showing torticollis without apparent cause are due to a slipped disc lesion.

3 Symptoms of a Slipped Disc Lesion between the Base of the Skull and C.1
When a disc lesion occurs at this site, admittedly rarely, the clinical picture shows a marked spasm of muscles on the affected side pulling the head slightly towards that side. This will give rise to pain over the

back of the head on one side or the other and sometimes even both sides may be involved.

The Thoracic (Dorsal) area of the spinal column consists of twelve vertebrae, the first dorsal articulating with the seventh cervical, the twelfth dorsal articulating with the first lumbar, in each case through the mutual annular ligament intervening. These two extremes of the dorsal region are the commonest sites of a slipped disc lesion of the dorsal area. One is closest to the neck where there is a great deal of movement and the other to the pelvis where there is a moderate range of movement. As in all disc lesions injury is the essential cause, which leads to a tear in the annular ligament and the prolapsing disc striking a nerve root. Acute pain will occur at the affected site with a moderate degree of muscle spasm, leading to a moderate degree of scoliosis at the affected level. Prolapsed (slipped) discs are not common between these two sites as there is very little movement of the vertebrae which are further stabilized by the articulation with the ribs. However they do occur and sudden injury plays a vital role. They are usually the results accidents such as a motor crash or a collapsed scrum in rugby."

SOME OTHER CAUSES OF BACK PAIN

Scoliosis

A condition that causes the spine to bend laterally. When viewed from the rear the spine can be seen to be bending either to the right or the left. It can be due to congenital deformities, paralysis of the muscles on one side of the spinal column, habitual bad posture due to repetitive work and the result of a slipped disc.

Facet Displacement (Subluxation)

We have learned that that each vertebra has seven bony outgrowths at its rear. Two are on top and two below. The vertebrae are arranged so that the top two of one vertebra mate with the bottom two of the vertebra immediately above. The bottom two of the same vertebra mate with a top two of the vertebra immediately below. These are the facet joints. Their surfaces, which are more or less in the vertical plane and covered in cartilage like the knuckles in a chicken bone, make fleeting contact in guiding the bending spine. Their other function is to limit spinal rotation. Injury can cause facet displacement; falling from a height and landing on one's feet for example. A facet may slip downwards on its opposing facet resulting in a bulging of the annular ligament on the affected side. If the bulging annular ligament strikes a nerve root the result is as if the disc had slipped.

Ligament Strains and Sprains
As you know, from the drawings of my simple experiment with the five children's building blocks representing the lumbar spine, running down the front and rear of the spinal column are two longitudinal ligaments which are attached to the vertebrae. (These are represented by strips cut from an old inner tube of a bicycle tyre.) The main functions of the longitudinal ligaments are to assist in maintaining or restoring erect posture and to protect the spinal cord from being damaged by a slipped disc. (See my simple experiment on pages 20 and 21) As the longitudinal ligaments are attached to the spine they are subjected to severe strains as the spine is subjected to daily activities that involve forward flexion of the body.

CHAPTER SEVEN

SITTING IS THE MAIN CAUSE OF BACK PAIN

"Squatting is the ideal way of resting your spine"

I have already mentioned the extremely important fact that the incidence of back pain among primitive tribes is extremely low. The reason for this is that they don't sit like you and me. They squat. Like babies. I'm absolutely certain that if, say, twenty infants were randomly selected and ten lived their lives sitting normally and the other ten had to squat and never to sit eight of the first ten would eventually experience back pain, whereas the "squatters" would be pain free.

"Squatting is the ideal way of resting your spine. In cultures where it's customary to squat backache is almost unknown." (Ref. The Back Doctor by Hamilton Hall FRCS(C)

I always ask back sufferers if they would agree that sitting with bad posture is the main cause of back pain. Ninety five percent will agree immediately. The remainder will do so hesitantly, after first saying their pain originally was caused by lifting, a car accident, a fall, the nature of their work, etcetera. Of course there are many occupations that are almost purpose designed to cause back pain: nursing, block laying, general construction work, farming, dentistry, the list is huge.

What is your opinion? Would you agree that sitting with bad posture is the main cause of back pain? Think about it for a couple of seconds. If the answer is "yes" then I'd like to ask you to answer three more questions.

Would you agree that no matter what the problem is in life there can be a number of solutions?

If the answer is "yes" I'd like to ask you another question.

Do you solve a problem by first applying the most obvious solution and if that isn't successful the second most obvious solution, then the third and so on?

If the answer is again "yes" I'd like you to carefully consider my last question.

So why don't you do this when back pain is the problem?

I know from experience you will have great difficulty answering this question. You have never previously given it serious consideration.

The very important point I am making is that you have already agreed the main cause of back pain is due to sitting with bad posture. Therefore good posture it is the most obvious solution to the problem of back pain. You have also just agreed that you solve problems by applying the most obvious solution. So why don't you do this with back pain? I would suggest you pause for a short time and consider the implications.

In all probability if you are reading this you are a back sufferer. You may be suffering on and off for years. Consider the various solutions you have tried, physiotherapy, chiropractic, osteopathy, surgery, GP's, painkillers, pain clinics and many more. In the majority of cases these treatments will have relieved your pain. Then off you go. Back to your old bad habits, continuing to abuse your spine minute by minute throughout the day. You are shocked when the pain reoccurs. And what do you do? Unthinkingly you go back for more treatment. And, as I've said before, the merry-go-round goes, well, round and round. But the interval between each "jaunt" gets less and less.

The vast majority of back sufferers hand over the solution to their problem to a third party. In most cases this approach may ease the symptoms. It is not tackling the cause. And so the pain <u>will</u> return. What they must do is get help from a third party and then eliminate the cause themselves. Nobody else can do this.

I regard this statement as absolutely essential to the permanent elimination of back pain. Unless you make up your mind here and now to change your lifestyle you will never be pain free. And remember the pain may get a great deal more excruciating and can be incapacitating. I've already mentioned one of my three essentials for pain free living which is:

You cannot hand over the solution to your back problem to a third party.

Unless it is due to non-mechanical reasons such as strained muscles, ligaments, or disease, it is my opinion most back problems exist for life. And as such will have to be managed on a constant basis. Doing this has become second nature to me. If I have to stand for a period I will find something to lean against. I avoid lifting and if this is not possible I will look for help. You hear on a constant basis the way the world has changed for the worse, huge increases in crime, self centered people not interested in helping others, etc. This may be so but I have found back pain is a great leveller. When I have to lift something I have no hesitation in asking a total stranger for help. Once I have told them I have a back problem they are only too willing to oblige. I can honestly say that no stranger has ever refused. As eighty percent of them would have suffered back pain most can empathize with my predicament and are only too willing to help.

You may feel too embarrassed or unwilling to admit your problem to strangers. Of course this is impossible if you haven't admitted it to yourself. I always tell back sufferers:

"The sooner an alcoholic admits to being one the sooner the problem can be controlled. The same applies to a back sufferer."

I've said before and I don't apologize for saying it again:

(1) When you are sitting, standing, lifting, with the "S" preserved there is the minimum possible stress on the spine.

(2) If you go through life with your spine at all times in the "S" position there is a huge chance you will be pain free. (Remember the primitive tribes.) I always keep the "S." And I am completely pain free. If I lapse, drive your car for example, I will get pain low down in my lumbar spine within thirty minutes, continue driving and it goes down my right leg. I can drive forever using my custom designed car seat mould.

My second essential for pain free living is therefore:

If you keep your spine in its S position at all times in the vast majority of cases you will be pain free.

An article in the August 2003 issue of "talkback," the magazine of BackCare (The UK National Back Pain Association) featured an interview with Mr. Andy Goldberg FRCS, Specialist Registrar in Orthopaedics, Royal National Orthopaedic Hospital, Stanmore, Middlesex. He is also a back sufferer. The article was in the form of questions and answers. I quote three of these:

Why are you interested in back pain?
"Being a long time sufferer of back pain is a great way of inspiring interest! What annoys me is that doctors, as a group, sometimes don't see the wood from the trees."

What do you mean?
"Throughout my life I have sought help from a multitude of disciplines. Doctors told me I had a bulging disc, physios that my pelvis was tilted and I had a sacroiliac joint problem, osteopaths said my back was "out," "in," curved, slipped or is it all coming from my sacrum? The language used by the community treating back pain is far too confusing. Therapies such as osteopathy and chiropractic as well as physiotherapy, pilates and others can help a huge number of people to manage their back pain, but there needs to be a much better integration of services. Until the various fragmented governing

bodies are able to talk the "same language" there will be no vast improvements for patients"

Why do you think back pain is so prevalent?
"Unfortunately, most people only pay attention to their backs when they are in pain and once this settles they stop looking after their backs, which is a big mistake - we need to address the underlying causes.

I think the most crucial issues relate to posture and lifestyle. We all sit badly, stand awfully in a queue, twist and turn in bed at night and pick things up without caring for the damage we are doing to our backs and the real problem is that poor posture doesn't damage you straight away. It can take years for it to show. Sitting or walking badly for an hour may not appear to do much harm, but do it for twenty years and that 's when the trouble begins. Hence why the prevalence of back pain is highest in the 30-40's age group."

I know I keep repeating that if S is preserved at all times in the vast majority of cases the result is pain free living. This may not happen immediately, just as "poor posture doesn't damage you straight away."(Andy Goldberg in the previous paragraph.) The following letter from an eighteen-year-old schoolgirl therefore gave me great satisfaction because of the immediate benefit she experienced.

July 1991

"Last March 1990 I severely injured my back while rock climbing. I have received no medical treatment and have been told it will be some time before I am relieved of pain.

I was in my fifth year last year and was unable to return to school until the beginning of the next school year. On my return I lay in the classroom for two months and continued to be unable to sit for any length of time.

As June approached I began to worry about how I was going to manage during my exams. On a trip into town one day, a week before

I was expecting to begin my Leaving Certificate I called into the Back Shop and asked was there a possibility of having a chair made for myself. After hearing my story they were only too eager to help, usually it takes up to six weeks to make one of these chairs.

On this occasion all those involved made every effort to have me measured and the chair made in less than one week, working straight through the June bank holiday weekend. I collected my chair completely upholstered and perfectly fitted on the day before my exams. For the two weeks of my exams for the first time in over a year I was able to sit without pain for more than an hour.

I am eternally grateful to all those working in and for The Back Shop. They made an extremely trying time for me a lot easier to deal with and my every day life more comfortable.

I wish to commend The Back Shop for all their help and support during my time of need."

The reason she was able to "sit without pain for more than an hour" was because she was working at the examination desk with her spine in "S" and her elbows resting properly on the armrests of the chair meant that a large amount of upper bodyweight was transferred from her spine down her arms, thereby greatly reducing the pressure on her discs.

This brings me nicely to my third essential for pain free living. Whatever your job or occupation is:

You must stop arranging your body around the work

CHAPTER EIGHT

HOW PERFECT POSTURE IS GUARANTEED

"A very flexible spine collapses into a worse posture than a more rigid one"

After years of development and a number of prototypes, I finally perfected my patented system, <u>Spinal System-S</u>, which provides the information to manufacture custom designed seating that guarantees perfect posture resulting in the relief of back pain.

The photograph shows the measuring chair in action. The person being measured was originally sitting with his spine slumped into a forward "C" position.

This is the position every unsupported spine adopts. Backprints are as unique as fingerprints and this certainly applies to spinal flexibility. A very flexible spine collapses into worse posture than a more rigid one.

During movement, running for example, a flexible spine acts like a spring and is better than a rigid one at absorbing forces generated by the feet striking the ground.

You can see the spine has been gently pushed into the "S"

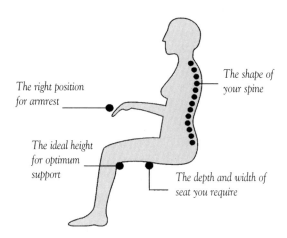

The right position for armrest

The shape of your spine

The ideal height for optimum support

The depth and width of seat you require

position. This is achieved in conjunction with the person being measured. Great care is taken not to apply too much lumbar support.

Measurements are also taken of arm heights. This is a very important consideration. Sitting with elbows unsupported results in all the upper bodyweight is being taken by the spine. When the elbows are correctly supported the result is that approximately 20% of bodyweight is taken from the spine and transferred down the arms, thereby relieving disc pressure.

The drawing indicates the various measurements taken in order to manufacture a home or office chair. (For a car seat mould only spinal curvature is measured.)

The seat height is important particularly for those who have difficulty getting from a seated to a standing position. The higher the seat is the easier it is to stand up.

The result of a back sufferer using a Spinal System-S office, home, or car seat is pain relief in the vast majority of cases. (80%-85% home, 90% office and, out of many thousands, there has only been a handful of failures with car seat moulds in the last twenty-five years. This is because it is difficult to do anything stupid as far as your back is concerned when driving. You are sitting back perfectly supported from the lumbar to cervical spine concentrating on the road ahead, so the car seat mould is always correctly used. Compare this with the home or the office where it is much easier to bend forward, slouch, twist, lean to one side etc.)

Spinal System-S products are also designed so you:

(1) Sit with your knees in line with your hips or marginally above. An exception would be somebody who has difficulty in getting from a sitting to a standing position. The higher the seat the easier it is to stand up. In this case the knees would in general be below the hip level. Back sufferers who don't have a particular problem in standing up nevertheless can put great stress on the spine in doing so. They bend the spine forward in the sitting position grasp the arms of the chair and literally haul themselves into an upright position. The best procedure is to squirm your bottom as far forward on the seat as possible, grasp the arms of the chair and using your arms push upwards, with your spine straight.

(2) In general, sit at an angle of 100 degrees (90 degrees + 10 degrees) between your thighs and spine, with the back perfectly supported from your bottom to your head.

The pressure on the intervertebral discs while lying flat on your back is the least possible. (Turn back to page 52 and examine that Swedish diagram again.) It is greatest when sitting with bad posture. The diagram also shows that sitting upright with good posture is still quite high, about half the pressure caused by sitting with bad posture. If you lay horizontally on your back on an electrically operated reclining chair and slowly started pressing the "up" button on the zapper, the pressure in your discs would increase in proportion to the distance your body rises upwards. That is until the body reached the vertical (90 degrees). After this point and the back of the chair continued to move forward, the angle would decrease below 90 degrees and there would be a dramatic increase in the disc pressure. This would be due to gravity taking over and the spine slowly slumping forward into the dreaded "C" position. So the more you sit back from the vertical the less the pressure in your discs. In my opinion setting the back of a chair at ten degrees back from the vertical is the optimum angle for most people. However the angle can be set to individual requirements.

In the last twenty years I have only encountered two cases, both women, where the person only got relief when they sat at an acute angle. In both cases the angle was 80 degrees.

I remember one of these women was a professor of anatomy. We spoke at length about a phenomenon I have been observing for many years and which was new to her. Most women can sit without moving for a very long time. I've seen them sit for three hours, almost transfixed, without as much as a tremor of a leg muscle. Men sit for a short time, something in the region of fifteen minutes, before the ritual begins. A slight involuntary twitching in the hips, a vague suspicion of a leg movement, a definite lateral hip movement, more of hesitant slide, then off they go into full-blown spinal contortion as they continuously twist and squirm, culminating in the most criminal of all spinal assaults. They move their bottoms forward in the seat, resulting in the spine abandoning all lumbar support. We failed to come up with any logical reason for this gender difference. She proposed the most likely one, "I guess men and women have different designed undercarriages."

(3) Arm heights are designed so that a considerable amount of your upper body weight is taken from your spine and transferred down your arms. Removing body weight reduces the pressure in the discs

If the arm heights are too low no weight is taken from the spine, too high results in raising the shoulders causing tension in the neck.

Resting both elbows keeps the body upright and prevents lateral flexion of the spine.

(4) The depth of seat is designed to match your physical specifications.

Seats that are too long means you can not sit back properly with your lumbar spine supported and also may result in the back of the knee being irritated, bringing about an involuntary forward movement of the bottom. This gets rid of the itch but eliminates any lumbar support.

An interesting fact is that most modern three-piece suites have seats that vary in depth from twenty inches to twenty-six inches. The average person would require eighteen and a half inches. So the shortest suite is an inch and a half too long, the longest seven and a half inches.

(5) The width of seat is designed to match your physical specifications.

The ideal width allows you to rest your elbows naturally on the two armrests and so transfer a large amount of your upper bodyweight from your spine and down your arms with resultant reduction in spinal disc pressure. Too wide and you have to lean to one side or the other to rest your elbows. This results in lateral flexion of the spine.

You must learn the characteristics of your spine:
I can stand for about forty-five minutes before my lumbar spine begins to act up: it's a feeling of tightness that slowly progress up into the thoracic, then into the cervical. If I stand unsupported for the whole of a football match the tightness will turn into pain. I go to all international matches in Landsdowne Road. A crowd of us meet up before games in the nearby Beggar's Bush pub and from there walk to the North Terrace (I prefer to watch standing in a crowd). Once I can find a crash barrier to lean on I am ok.

I can sit having a meal on a dining room / kitchen chair for maximum of twenty minutes without pain. When I use a Posturite, a special shaped lumbar support cushion I designed made from viscoelastic (pressure relieving material designed for space travel that softens with body heat and moulds itself around the lumbar spine), I can sit for several hours. (www.back-shop.com for details) The same applies wherever I sit. Knowing this if I travel by bus, taxi, train, airplane and particularly a strange car, I use a Posturite.

I have no problem playing golf. This is because a golf swing is a repetitive movement and it either causes pain or it doesn't. I'm lucky in this respect. If you played golf with me for the first time you might think that as I stood on the first tee I was starting to pray. I take this little plastic tee from my pocket, kneel down on my right knee, insert the tee into the ground and place the golf ball on top. Bending, even momentarily, causes me pain. Whenever I get the ball into the hole I retrieve it by means of a rubber sucker attached to the handle of the golf putter. You do more damage to your spine by bending down on a golf tee than in a four-hour game of golf. And remember there are eighteen tees on most golf courses.

I can walk forever. Several times over the last twenty years when I was in pain I forced myself to walk. Gradually I started to get relief and after thirty minutes was probably eighty percent better. Walking as briskly as the pain will allow, swinging the arms, head up results, for me, in moving whatever was causing the pain out of contact with the nerve.

Nineteen years ago I had a premises in Fitzwilliam Square, it was winter, bitterly cold, wind and sleet, my back pain excruciating. My wife Chris and my daughter Tara called. I didn't want them to know the extent of my pain and volunteered to walk to Baggot Street some three hundred yards away for cream doughnuts. The pain was so great I had to give them to Tara to carry. If you are carrying a weight the stress on the spine is not just the weight. It's the weight multiplied by the distance from the spine. So if you must carry a weight hold it as close as possible to your body. When we reached my premises I went down the sleet covered wooden steps to the basement, slipped, tumbled like a rag doll, slammed onto the concrete yard, one leg hooking around the bottom banister and doing a passable imitation of a leggy "Moulin Rouge trooper" doing the splits. When I eventually, fearfully, got to my feet, miracle of miracles, I was pain free. The extreme jarring had moved the "foreign body" out of contact with the nerve. This was the last time I experienced chronic pain. However I would not recommend you try this remedy.

While walking briskly can be beneficial, running is to be avoided. When walking you have one foot in contact with the ground at all times. So the forces compressing the discs are relatively small. Running by its nature means pressing down with one foot, causing the body to leave the ground, coming down hard on the other foot, pressing down and so on. Landing on the ground, particularly hard ground, results in large continuous compressive forces being applied to the discs. My spine would not be able to withstand this continuous jarring.

The mention of "walking" in the previous paragraph brings to mind "shoes," which people rarely think of in relation to keeping their backs healthy. I am referring to shoes for daily living. As with choosing chairs for the home, shoes are considered articles of fashion and most people buy them because of their appearance. Once they comply with this criterion, are the correct size and not too tight, a quick walk up and down the shoe shop and they are purchased.

It is my opinion that shoes should be designed in such a way that the foot strikes the ground in exactly the same natural way as when no shoes are worn.

Examine how people walk in their bare feet. The heel strikes the ground first, then the foot rocks downward on to the ball of the foot, bringing the heel away from the ground. The foot continues to rock until the area between the ball of the foot and the big toe is in contact with the ground. Then downward pressure, angled backwards, drives the body forwards until the other foot strikes the ground and so the action is repeated.

When standing in correctly designed shoes the feet should be horizontal to the ground, exactly as when standing in bare feet. Many sports and walking shoes achieve this.

The average shoe has a heel that is about two centimeters thicker than the sole. As this is something less than an inch you might think it is not worth worrying about. But it is. The result of the heel being higher than the sole is that when the heel strikes the ground the foot has to rock downwards an extra two centimeters more than when bare footed. This results in the natural movement, whereby the heel strikes the ground and the bodyweight is immediately transferred in a smooth rocking movement on the ball of the foot, being destroyed. The ball of the foot travels something in the region of two centimeters below the horizontal and literally hits the ground with a slap. Try walking on linoleum and listen to the noisy slaps as the sole of your shoe hits the surface. These send continuous shock waves up your feet into the ankles, up your leg to the knees and hips and finally into the body's suspension system, the spine.

Most car drivers are aware of the vibrations that are felt in the steering wheel as a car is driven over cobblestones. Pedestrians can hear the clump-clump-clump of the tyres moving over the surface. The vibrations are tiny and don't cause any harm to the car's suspension. However if the car were to spend its working life, say 150,000 miles, running exclusively on cobble stoned roads would this make a difference? Of course it would. The coiled spring in the suspension might be suffering from metal fatigue and the hydraulic damper might be leaking. The same thing could happen to your spine as a result of years of wearing badly designed shoes. I don't know of any scientific research in this area. But it's a simple thing to rectify. Are you willing to take a chance and ignore it?

To illustrate what I am talking about try this simple experiment. Take off one shoe. Measure the thickness of the heel and the thickness of the sole. Say the difference is two centimetres. Look around you. Find something flat that is approximately this thickness, a paperback book, a wad of paper, whatever. Suppose it's a paperback book. Tie the book to the underneath of the sole with an elastic band or a piece of twine. Put the shoe back on and walk around the room. You will immediately be aware of the huge difference. The foot in the shoe with the book attached touches the floor smoothly without vibration. The other shoe hits the floor with a slap; you can actually feel the shock waves.

Another consequence of shoes with heels higher than soles is to move the spine slightly forward of its natural "S." This is further compounded by shoes that are designed, with fashion in mind, with the sole of the shoe bending upwards under the toes. The foot has now to rock even more forward until it achieves "lift off" as the big toe pushes it off the ground.

I would never, ever, attempt to play games like tennis or squash, where sudden involuntary movements are used: bending down, short sprinting, sudden stopping, reaching up, reaching back, twisting, hitting the ball coming from a multitude of different angles throughout the course of a game. Even writing this makes me wince.

I am a reasonably keen gardener and can do this indefinitely without abusing my spine.

A spade and fork are two of the basic tools used by most gardeners. The most popular ones are a modern design, waist high, with polished wooden handgrips. Avoid these like the plague. They put huge stress on the spine. The procedure is to press the spade or fork into the ground with your foot, then bending forward with one hand holding the handgrip and the other the handle levering the soil out of the ground by brute force. Think for a moment what this is doing to your spine. I'd last probably fifteen minutes before giving up in agony. Use the old-fashioned type spade and fork with a long straight handle. Operate by pressing into the ground with your foot in the normal way.

Then, all the time standing straight, pull the handle back towards you. This breaks up the earth. If you want to turn the sod merely rotate the handle with your two hands. When weeding or planting always use kneepads. Go down on both knees. Bend from the waist with a straight back, support yourself by placing one hand on the ground in front of you, then use the other hand to do the planting, weeding, etc.

The foregoing will give you an appreciation of some of my spinal characteristics. Give some thought as to what yours are. Consider how they can be overcome and plan to do so. This is the first step in taking responsibility for solving your problem and not delegating it to a third party. By all means seek help from a third person, a chiropractor, a physiotherapist, an osteopath, etc. I can't emphasize strongly enough that this is "help."

After a course of treatment in most cases you will feel better. The rest is up to you. If you have to return for more treatment the cause is probably due to spinal abuse.

CHAPTER NINE

SPINAL SYSTEM-LIFE

"Spinal System-Life = my three essentials for pain free living"

I am going to take you through a day in my life in order to demonstrate how it is possible to live with the spine never moving out of its "S" position. I call this my "Spinal System-Life."

In essence I am copying how modern day primitive tribes keep their spines at all times in "S." They never sit; they squat. Of course I don't squat. However I make sure the pressure in my discs is similar to that when squatting by always sitting with "S." When not sitting I work hard at never allowing my spine to slump or bend into a forward "C." Having done this for many years it is now second nature, requiring little conscious thought.

My Spinal System-Life results in me leading a mainly unrestricted, pain free, life. When I explain briefly that this is due to sticking firmly to my straightforward system of always keeping "S" most people immediately say this is impossible, or else are dismissive of the idea and make no comment. In order to prove this is possible I'm going to take you through a day in my life. I'll start off in the morning and chart my progress, "spinewise," through the day. I will vary this with comments on what other people may be doing. I think it would be of great benefit to you if, after you have read what I am about to write, you did a similar analysis of an average day in your life. At the end of this chapter I suggest a simple method to help you achieve this objective.

I am starting by examining beds.

Many experts dealing with back pain have a fixation with the word "orthopaedic." I am referring to orthopaedic beds and orthopaedic chairs. I constantly come across back sufferers who have been advised to get one, or both. An average of three customers per week asks me for an orthopaedic bed or chair. When I ask them why, they say the doctor advised them. For the last twenty-five years I have been speaking out strongly against this advice

The word "orthopaedic" means "the branch of medicine dealing with the treatment of disorders of the bones and joints and with the correction of deformities, originally in children."(Ref. Oxford Dictionary.) The word "orthopaedic" when used as an adjective to qualify a "bed" or "chair" means hard, like as a rock. The only logical reason I can come up with for the attachment of this word to two household items is that maybe in the past orthopaedic surgeons told patients to use hard beds and chairs and then over time the name of their profession became synonymous with hard beds and chairs. I remember when I was in hospital the bed not only fulfilled all the requirements of the modern interpretation of the word "orthopaedic," but had planks placed under the mattress for good measure.

When you lie on an orthopaedic mattress on your back there is no give in the mattress under your bottom. It remains absolutely level with no depression. The same applies under your shoulder blades. The lumbar spine being unsupported collapses into a horizontal C resulting in pain.

I'm blessed as far as sleep is concerned. I could lie down on the floor at any time, night or day and will be asleep in a couple of minutes. The same would apply if I slept on an orthopaedic mattress. And like lying on the floor I would be awake in a very short time. It is impossible for me to get a night's sleep on an orthopaedic mattress. I would spend all the time tossing and turning, waking up, going back to sleep, aching, waking up again. I think you've got the picture. I admit that every now and then I come across somebody who finds an orthopaedic mattress works. My advice is, as it is with everything in life, "if it works stay with it."

I have been advising people against the use of orthopaedic mattresses for last twenty-five years. When I hear that a person is using one I advise a visit to a filling station to buy a gallon of petrol and to set fire to the mattress in the back garden. I know this procedure would not be environmentally friendly so perhaps they should dispose of it in a way that complies with byelaws.

This is Walker-Naddell FRCS on the subject (The Slipped Disc and the Aching Back of Man):

"Many patients are advised to lie on a hard bed for 6-8 weeks. Bedding manufacturers have produced these so-called "orthopaedic beds" to simulate the hardness and flatness attained by placing a sheet of hardboard under a firm mattress. I personally deplore this form of therapy and consider the term "orthopaedic" wrongly applied to these beds. The term "orthopaedic" simply means the correction of a deformity and in my opinion the orthopaedic bed not only does not correct, but actually aggravates the condition.

Anatomically the body is not designed to lie on a flat surface. Even in the army I noticed young and fit soldiers asking for help to get up after sleeping on the hard ground all night. Thus if perfectly healthy young men find sleeping on a hard surface intolerable, it seems inconceivable that similar conditions would prove beneficial to a back sufferer. My experience has been that, in almost every case of a painful back, lying on a hard surface has only aggravated the condition.

Most back patients prefer to sleep on their sides or back. If the patient lies on his side on a hard bed, the bony prominences of the shoulder or hip bone are held fixed and the remainder of the lumbar area extending up towards the dorsal aspect of the spine, sags unsupported downwards to the hard surface. If, however, the patient lies on a good interior sprung mattress the bony prominences of the shoulder and hip area are impaled into the softness of the bed and between these points the mattress forms a bulge which gives support for the affected lumbar area of the spinal column. Similarly, when the patient lies on his back on a hard flat bed, the posterior aspect of the upper dorsal, i.e. the scapular region, forms one fixed point and the posterior aspect of the pelvis forms the other. Thus the area of the spinal column between these two points of the back is left unsupported."

I was gratified to read in The Times (of London), dated 14-11-2003, that The Lancet journal, the authoritative voice of the medical profession, had recently published the results of a Spanish study that verified what I had been saying. The research turned the traditional medical wisdom that recommended that hard mattresses help to alleviate back pain, on its head.

Never buy a mattress without trying it out first. I know this can be difficult, as ideally you would have to sleep on a mattress for several nights before you could be certain it is the one for you. The criterion when buying a mattress is to lie on it. Your bottom should make a depression in the mattress, so should your shoulder blades. This results in the part of the mattress between your bottom and shoulder blades supporting your spine. The same applies when you sleep on your side. Your shoulders and hips dig in and your spine is supported latterly.

Study the two drawings.

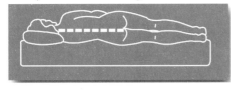

Lying on your side your hip and shoulder dig in, resulting in lateral support for your spine. See how your spine is absolutely straight.

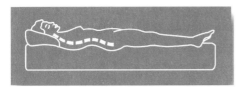

Lying on your back see how your bottom and shoulder blades cause depressions in the mattress and how this ensures the lumbar spine is supported and the complete spine is in its "S" position.

People ask me which is better, to sleep on your back or your side. As you are the only person who knows your own spinal characteristics, you are the one best able to answer this. I can sleep in both positions, so I do so. If a particular sleeping position causes you a problem, then don't lie that way. (The Swedish study we have examined shows that the pressure in the discs is lower lying on your back than on your side.)

For many years I supplied a conventional spring mattress that fulfilled the above criterion with good success and used one myself. In spite of this my wife had been complaining that when I got out of bed every morning the bed looked as if a herd of elephants had been trampling on it. I had spent the night, not in pain, but aching after about ten minutes in the one position. I was not conscious of turning but obviously I did this on a continuous basis to relieve the aches, hence the rumpled bedclothes.

We spend one third of our lives in bed, so it is extremely important the mattress supports the spine in S. As you know lying results in the lowest pressure in the intervertebral discs. If you wake in the morning after a night's sleep there is almost certainly a lack of support in the mattress.

A simple way to check if your mattress is providing proper support is to lie on your back on the mattress. Try to slide the palm of your open hand between the mattress and your lumbar spine. If it slides in easily there is insufficient support. If you have difficulty there is probably good support and your spine will be in a horizontal S and not a horizontal C.

For the last three years I have been lying on The Back Shop Memory Mattress made from visco-elastic and when I get out of bed every morning the bedclothes are in pristine condition, the result of hardly moving position throughout the night.

Visco-elastic is a revolutionary new material and is the ultimate in mattress technology.

The brochure reproduced on the following pages describes in detail how the Memory Mattress works.

THE BACK SHOP
MEMORY MATTRESS

My name is Colm Campbell, an engineer, who had a serious back problem. I have been running The Back Shop in Exchequer Street for over twenty years.

I am now totally pain free, thanks to my invention, Spinal System-S, which gives the information to manufacture custom made office, home and car seats that guarantee perfect posture. Some ten thousand back sufferers have used the System-S, and are now sitting without pain.

I am pain free because at no time during the day does my spine go out of its own unique S shape. Lying in bed has always been my problem. I have spent many years experimenting with various types of mattresses, trying without success to find one that would give pain free sleep, with no aches in my joints the next morning.

Over the years no matter what mattress I used my wife complained that, when I got out of bed, the bedclothes looked as if a herd of elephants had been trampling over them, the result of continuous tossing and turning in a vain effort to get rid of niggling back pain.

I solved my problem some five years ago by placing a two-inch layer of pressure relieving visco-elastic on top of the mattress. Visco-elastic was invented by NASA to relieve the incredible G-forces experienced by astronauts during take-off.

Since then lying in bed has been bliss, and when I get out of bed every morning the bedclothes are in pristine condition, the result of hardly moving my body position throughout the night. The combination of visco-elastic and the conventional spring mattress results in my spine always being in its correct sleeping posture. The disadvantage is that the two inch layer of visco-elastic creeps and has to be adjusted back into position before a night's sleep. We have now solved this problem.

The Back Shop Memory Mattress (see above drawing) duplicates the sleeping system I perfected over the last five years.

The mattress consists of two layers. The top three-inch layer is visco- elastic, the bottom a six-inch layer of special springs, which prevent rolling into the center. This combination results in perfect posture, and the feeling of the body being suspended in space.

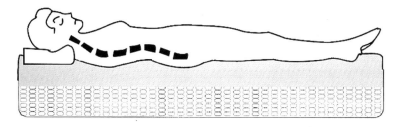

One of the characteristics of visco-elastic is that it is temperature sensitive, and softens with heat. Lying on your back, or side, results in the top layer of the mattress slowly attaining body heat, softening, allowing the body to sink into the mattress, until eventually it takes up an exact imprint of your body. Your bottom and shoulder blades slowly sink into the mattress thereby guaranteeing correct support for the lumbar and upper spine.

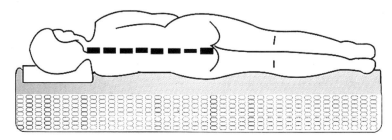

Similarly, lying on your side, the hips and shoulders sink in, giving lateral support to spine.

Moving position results in the mattress immediately starting to lose the imprint of your body and coming back to its original shape. It remembers where it should be; hence we call it " The Memory Mattress."

Another characteristic of visco-elastic is that it immediately collapses when pressure is applied. This can be demonstrated by placing a large bunch of keys on the mattress. Sitting on the keys causes the visco-elastic underneath to immediately collapse, resulting in the keys sinking into the mattress, and the person being totally unaware of the keys being sat on.

The practical application of this is that where bony parts of the body (ankles, heels, knees, hips, elbows, shoulders, shoulder blades) touch the mattress it immediately collapses. Pressure, the cause of bedsores, aches and pains etc., is removed from the joints and distributed over adjoining areas.

The combination of visco-elastic and coiled springs creates "bounce", allowing the body to move position more easily.

The Back Shop Memory Matress is the best mattress I have ever slept on. It is available in most sizes. You are welcome to call to the shop and try it for yourself.

We are so confident of this claim that we offer a no quibble ten-day money back guarantee.

The Back Shop Memory Pillow is made from the same visco-elastic as the mattress. It also softens with body heat, moulds itself around your neck, resulting in the natural curve of the neck being correctly supported throughout a night's sleep.

The following extract from a letter typifies many responses;

"For the first time in months, since I had back surgery, I am at last getting a full night's sleep. This mattress gives my back support at all times, resulting in a great feeling of comfort. Before that I found each mattress either too soft or too hard, resulting in restless painful nights. Now with the Memory Mattress, not alone do I get a good night's sleep, but also my back is vastly improved. In fact it is only when away from home and sleeping on another mattress I realize just how good and how much support to my spine the Memory Mattress gives."

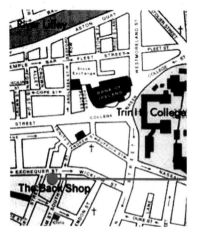

GETTING OUT OF BED

Lying flat on my back I move my left leg to the left, until the calf and foot are over the edge of the bed. (The option is to move your right leg to the right if you sleep on the right side of the bed.) I then brace my calf and foot against the bed. I roll my body to the left and, with my left elbow resting on the side of the bed, continue to roll until my legs are clear of the bed and simultaneously press myself up with my left hand. I am now in the seated position with my feet on the floor. This may sound complicated, but it's really simple and needs very little effort. You have now got from the lying position to the sitting without bending your spine, which is still in its "S" position.

I always use slippers without backs, so I can easily get my feet into them without having to bend. I then go into the shower room where the sink and shelf holding my shaving kit etc., are sufficiently high to prevent me having to bend.

I then have a shower. I never have a bath. Climbing in and out will create great pressure in my discs. This could result in pain. Sitting in a bath for even a few minutes will result in my spine going from S to C and will cause pain. One of the characteristics of the spine is that damaging the spine may not immediately result in pain. It can take hours, days and even weeks to manifest itself. The outcome is the back sufferer may not remember the cause when the consequences suddenly strike. Of course one of the main points I am making in this book is that the vast majority of people unknowingly damage their spine on a constant basis throughout the day. They also do this at nighttime if their mattress is not providing horizontal S.

After the shower I dry myself with a large bath towel. The big problem for me then is how to dry my feet without bending. I have solved the problem very easily by draping one end of the towel on the floor. I put one foot on the towel. I then drape some of the towel over the foot and using the other foot like a spare hand, I dry the foot underneath the towel.

I am back in the bedroom and now the biggest problem for me and one I have had since I hurt my back playing rugby when I was sixteen years old, arises. How to put my socks on? What I do is go down on one knee, extend a foot slightly forward and with a great deal of effort put the sock on. I move the knee of the foot I am putting the sock on sideways and fit the sock from the instep side of the foot. I then drag the foot forward along the floor resulting in the sock encasing the whole of the foot. It may appear complicated, but it's not. I repeat the procedure for the other foot.

I use socks that are a size larger as they are easier to get on. Putting on my shoes is no great problem. I always leave them with the laces as loose as possible and so can push my feet into the shoes while standing upright. I solved the problem of the laces coming out of the eyelets years ago. I tie a knot at the two ends of each shoelace, so they can't come out of the eyelets.

In the kitchen, having breakfast, I sit at the table using my Posturite lumbar support, which is permanently on my chair.

We have a number of vans and one of these picks me up every morning to take me to work in Exchequer Street. Again, I use a Posturite lumbar support. I rarely travel to work by car, which is fitted with a custom designed Car Seat Mould.

During the working day I spend a lot of time on my feet, talking to back sufferers, advising them on back pain, its causes and prevention If I have to stand for a lengthy period I lean against any available support. I always kneel when measuring customers' spines. The only time I sit at a desk is when sending and receiving e-mails, writing letters and checking the number of hits on my website for the previous day. (I get a lot of satisfaction that a simple idea I had twenty-five years ago has now its own website and keying two words "back shop" into Yahoo reveals that it varies between first and third position from amongst millions catalogued under the same heading.)

I now want to write in detail about a huge problem and of all the things that cause back pain throughout the world, this is one of the main culprits. In my experience ninety five percent of all people who work at a desk do so with the most atrocious posture possible.

Just look around you, look at yourself, look at people on television; almost everybody sits hunched over their desks. An interesting fact is that probably ninety nine percent of the many thousands of men I have met, who work at a computer, cannot touch type and spend their day picking, hen- like, at the keyboard. Almost every woman doing a similar job can touch type. If I had anything to do with the education of very young children I would insist that when infants are being taught the alphabet and are learning to recognize and write letters they should simultaneously be introduced to the keyboard.

WORKING AT A DESK

**"Good posture is the most important way of preventing back pain."
(Ref: Back Care by The Health Education Bureau.)**

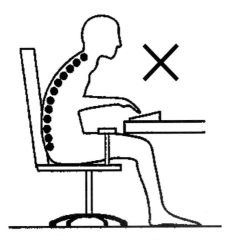

Nowhere is this more applicable than in an office situation. While no statistical information exists, the vast majority of people working in an office sit bent over the desk. As you know this "C" position creates great stress in the lumbar and cervical spine, with resultant huge pressure in the discs and eventually pain.

Consider how you sit. Look at other people. You will realize the truth of this statement.

Spinal System-S office chairs guarantee perfect posture. By following some simple procedures, the result in the vast majority of cases, is pain free sitting throughout the working day.

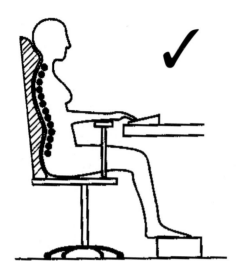

These procedures are:

Use a footrest. Most desks are approximately twenty eight inches high. This means people of average height have to sit with their knees well below their hips, resulting in the lower body tending to pull the upper body forward and thereby helping to bend the spine into the "C" position. A footrest (four inches high for the average person) raises the knees causing the spine to tilt backwards to a better posture.

Sit as close to the desk as possible. This is important, as it is another aid to preventing the "C" position. It does away with having to bend forward to pick up the telephone, working documents, etc.

Adjust the height of the seat so that, with the elbows resting on the armrests, the forearms are gently sloping down to the keyboard / mouse. Avoid too steep an inclination. Correctly resting the elbows results in a large amount of upper bodyweight being taken from the spine and transferred down the arms.

Work at all times with the spine in its "S" position, fully supported from the cervical to the lumbar spine and with the elbows supported. Working with the head supported removes all stress from the neck. People who must look down at the keyboard should support the spine as least as high as the shoulder blades.

For hand writing / reading use a simple portable lectern, sloping from approximately four inches at the rear to a half inch at the front. The objective is to bring the work up to the person.

I don't sit at a desk for long intervals of time during working hours. But I can do so at home when I sit, feet on a four-inch footrest, with

Severe spinal damage at 0mph.

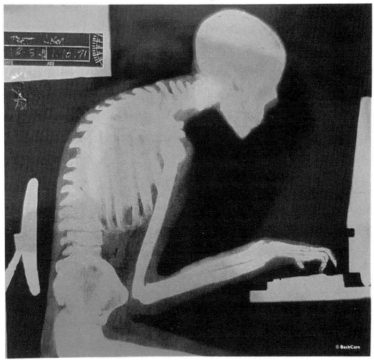

Poor posture can result in back pain;
check your posture, take regular breaks, and get
an ergonomic assessment of your workstation.

Contact: BackCare, 16 Elmtree Road, Teddington,
Middlesex TW11 8ST. Phone: 020 8977 5474
e-mail: back_pain@compuserve.com website: www.backpain.org

BackCare, the new name for the National Back Pain Association,
is a Registered Charity (No. 256751)

The Back Shop is a corporate member of BackCare
(The U.K. National Back Pain Association)

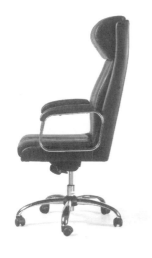

the chair tilted backwards as far back as possible, an angle of approximately 45 degrees.

Remember we have established that the more the person in the seated position tilts the chair backwards the less the pressure in the discs. I could sit down right now and work in this reclined position for the next eight hours with hardly any breaks and I will be completely pain free. Furthermore the action of standing up, when I have finished working will not cause even a tiny twinge. Of course not everybody can work in this position because the type of work they do may demand that they sit more upright, working with accounts for example. Also you must have touch-typing, which as I have said already rules out most men.

If you are one of these I consider it is absolutely essential that you learn to touch type. Would you buy a car, didn't learn the layout of the gears and every time you changed gear meant you had to look down at the diagram on the gear stick? Of course you wouldn't and I'm not referring to the fact that this would constitute dangerous driving. You would be embarrassed that the word might get around that you were unable to master a simple mechanical task and your ego would be affected because you could not operate a piece of equipment, the car, at the optimum. Anyway part of the enjoyment of motoring is zipping up and down the gears. This attitude certainly doesn't carry over into touch-typing that, for me, gives similar satisfaction. Most men are rather boastful when they admit they can't touch type. "Didn't do it in school," is the stock reply. What they are saying is they went to a boy's school. Typing is for women. There are a lot of things they do now that they didn't learn in school, but I never pursue this stupid statement.

Ten years ago I did three nights, 6.0 pm to 9.0 pm, of a short term typing course. Because I took no breaks during each three-hour stint I suffered quite a lot of pain in my hands and wrists, the result of using dormant muscles. I never got to the bottom row of letters. At the end of the three nights the teacher scored me high for determination, but zilch for accuracy. However I continued to practise at home and gradually after several weeks was able to type, admittedly very slowly. Now touch-typing is second nature.

DRIVING

I am going to talk about driving at this point because many people do it for a living, it is part of our daily lives and would rank on a par with working at an office desk as a major cause of back pain.

Most car seats appear to be almost fiendishly designed to cause back pain. I can drive any car pain free for approximately fifteen minutes when I begin to experience pain low down in my lumbar spine. After about forty-five minutes the pain will have increased hugely and I will be experiencing pain down my leg. Anything over an hour and I will be in excruciating pain.

There is no connection between badly designed seats and the cost of a car. Some of the worst designed seats belong to two of the most prestigious makes of cars. I mentioned these by name on my website a few years ago. However on legal advice I withdrew them.

Many seats, including these two, are well engineered with all sorts of clever adjustments; yet totally fail in the main function of supporting the spine in the "S" position. Badly designed car seats allow manufacturers to design cars with low sleek rooflines. When customers of around six feet in height and over come to me for a car seat mould I ask them to sit in the driving seat, arch their lumbar spine forward, place their balled fist in contact with the lumbar spine and lean back. In most cases their heads will touch the roof. The spine is a flexible mechanism. Pushing the lumbar spine forward (flexing) results in raising the head upwards and backwards by about one and a half inches, enough for tall people to make contact with the roof. Try this yourself. You'll see what I mean. Therefore if cars were designed with seats that provided even good posture the rooflines would have to be higher, thereby destroying the esthetic look. Several years ago I approached a major manufacturer with a design that would allow, by unzipping the back and twiddling a few knobs, a car seat to be custom designed to provide "S" within a period of a few minutes. They told me they weren't interested.

The Spinal System-S Car Seat Mould is custom designed to convert your badly designed car seat into one that provides "S."

As shown the Car Seat Mould fits on to the existing seat. It is clipped into position by a strap that passes under the headrest and is buckled to another one that comes from the bottom of the seat. It can be moved from car to car.

As you can see you are supported from the head down to the lumbar spine. In the event of an accident whiplash is eliminated.

Many people are aware of the sharp outbreak of pain that can occur during a journey, becoming increasingly excruciating as the journey continues, resulting at journey's end with even more pain caused by rising from a sitting to a standing position.

The Car Seat Mould results in endless hours of pain free driving. Out of many thousands I have had only seven or eight failures in the last twenty-five years. I define failure as not being able to drive four hours pain free. The reason why car seat moulds are so successful is because it is

Motorway Madness

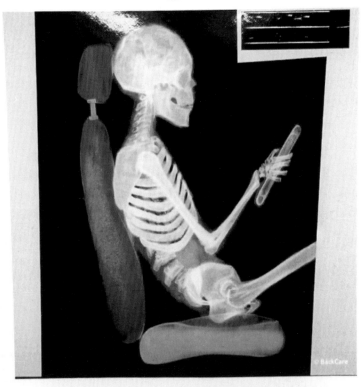

Bad seating, long journeys, no rest stops and tension
can all contribute to back pain.

To reduce the risk, adjust your seat properly, take
plenty of breaks and talk to us.

BackCare, 16 Elmtree Road, Teddington,
Middlesex TW11 8ST. Phone. 020 8977 5474
e-mail. back_pain@compuserve.com
www.backpain.org
Registered Charity No. 256751

**The Back Shop is a corporate member of BackCare
(The U.K. National Back Pain Association)**

difficult to abuse the spine while driving. You sit with perfect posture and, with your hands on the steering wheel, the dreaded C position is difficult to attain. The Car Seat Mould is covered in a material of the driver's choice to match the existing car seat.

Many people drive in the upright position. When asked why, they say, "the doctor told me."

Sitting upright, which I define as vertically, means the spine will slump into a "C" position. In twenty-five years and out of something in the region of ten thousand back sufferers, I have encountered only two people, both women, who could sit for long periods with the spine in the vertical. Try this for yourself. Right now. Feel the tension; down low in the lumbar spine. After a couple of minutes you will become most uncomfortable. Relax and the spine will slump forward, giving momentary relief. However your spine is now in C and you know the consequences of this. Extremely strong well-toned stomach muscles are necessary to keep the spine in the vertical. Although I never did it myself, I would advise people, if they have the inclination, to pursue a course of exercise designed to strengthen these muscles that can play a very important part in always achieving S. The method I am proposing for always preserving S is simple and can be used by everybody. A muscle-building course requires more dedication.

A huge number of car seats are designed so that the driver is sitting with the knees much higher than the knees. This fact, combined with the forward "C" position, can result in the angle between the thighs and the spine being something in the region of seventy degrees. This results in huge pressure in the discs. If it were possible to drive flat on your back and steer by means of a series of overhead mirrors, the pressure in your discs would be the least possible. I would recommend at least ten degrees back from the vertical. I normally set the back of my seat at around fifteen degrees. When I make a car seat mould for very tall people in many cases they have to sit at this angle, more upright would mean their heads would be touching the roof.

An illustration of the effect of a badly designed seat, with the knees much higher than the hips, happened to me about six years ago. It was a cold, drizzly, inky black, winter's night. I was taking my daughter Tara horse riding in Celbridge. The car was an old Mini, complete with automatic gearbox. We were driving along the dark narrow road, chatting, then bang, without the slightest inkling of what was about to hit me, my right leg felt as if it had been assaulted by a chain saw being operated at full revs. I have experienced pain, but this was on a different level, the worst of all time. I swung right into the entrance of Weston Aerodrome, jammed on both handbrake and footbrakes and was literally out of the car before it had come to a standstill, all thoughts of "S" completely forgotten. Some five minutes later the pain had subsided to a large degree, but I was reluctant to get back into the car knowing the possible consequences. However I couldn't stay standing in the darkness forever, so I gingerly got back in. I sat down and within seconds the pain disappeared. While I had been doing my frenzied "hop scotch act" Tara, without saying a word, had taken off her overcoat, rolled it up and placed it at the back of the driving seat so that when I sat back in my knees were level with my hips.

There are two types of manual adjusters for the backs of car seats, one a lever, the other a hand wheel. Avoid the lever type. It allows only a small number of adjustments to the rake of the seat. A hand wheel gives a large number of tiny adjustments. These fine adjustments are not available with the lever type of adjuster which allows a large amount of travel from one setting to the next. On long journeys adjust the rake of the seat at regular intervals. Every time you move your spine forward or backward, even by a small amount, you will find it may relieve any nagging ache. This is due to the fact that even slight movements can transfer pressure from a part of the spine that is aching to an adjacent area and so relieve aching.

When I get into a car I go in head first with an S spine. I rest my left hand on the hand brake lever, or whatever is between the seats, hold the top of the steering wheel with my right and simply drop into the seat, still with "S." Getting out, I swivel my bottom to the right until my feet are on the ground. I reach up with my left hand and take hold of the top of the steering wheel, grasp the top of the door with my right and by a combination of pushing upwards with my legs and pulling up with my hands I get to the standing position, without bending.

I get home from work around six o'clock and have an evening meal. I would never, ever, sit at a table without a Posturite lumbar support which, as I have told you already, is permanently on my chair. Without it I would be in pain in a very short time and I use it everywhere.. I did slip up a few years ago. I was so excited going to the National Concert Hall to listen to Neil Armstrong, the first man on the Moon, that I forgot to bring the cushion. Within twenty minutes I was in great pain. I stood up, took off my jacket, rolled it in a ball, jammed it into my lumbar spine and spent the next three hours in pain free wonderment at the achievements of this unassuming, witty, dare I say "down to earth?" astronaut.

After an evening meal I normally crash out for several hours watching television. This brings me to the third in the triumvirate of major causes of back pain, the office chair, the car seat and now the dreaded lounge chair.

LOUNGE CHAIRS

Everything negative I have previously said about office chairs and car seats applies in equal measure to lounge chairs.

The one universal criterion the vast majority of the world's population applies to the problem of choosing a lounge chair is "how does the chair look." They will sit in whatever chair they like the look of bounce up and down a bit and do a sort of combined gyrating movement of bottom, hips and shoulder blades. They say to whoever is with them, "Ooh! This is comfortable." They equate comfort with softness; the big squishy chairs allow the complete human frame to collapse totally out of alignment, like a half filled bag of spuds chucked on to the side of the road. The excited would-be buyer doesn't realize that "Even though bad posture may not cause any discomfort, continual poor posture <u>will</u> in the long term cause back pain." (Ref. Back Care by The Health Education Bureau.) Cast your mind back to the last chair you bought. What criterion did you use when making your final choice?

As with office chairs and car seat moulds, the Spinal System-S range of home chairs guarantee "S." The result in about 85% of cases is pain free sitting.

The spine is pushed into its own unique "S" shape. A template is

made from the shape and from this the back of the chair is made to precisely reciprocate the person's backprint and so guarantees "S."

The heights of the arms are made so that, with the elbows resting, a large proportion of the upper bodyweight is taken from the spine and transferred down the arms, thereby reducing the pressure in the discs.

The depth of the seat and the width of the seat, is made to individual require-ments.

In general the best sitting height is when the knees are in line with the hips or marginally above. But nothing is constant. Somebody who has difficulty getting from a sitting to a standing position would need to sit

with the knees below the hips. The higher you sit, the easier it is to stand up.

The best type of chair is a reclining chair. When you are sitting for some time and you develop an ache in your back the inclination is to move from buttock to buttock to get relief. The worst thing is to move your bottom forward and so eliminate any lumbar support. When you get these urges in a reclining chair you simply recline the chair, thereby getting relief.

I have sat watching television on a reclining chair three times in the last year from twelve noon on a Sunday until midnight without pain. The last time was two soccer matches, an F1 Grand Prix and Tiger Woods playing golf.

As you can see the chairs look normal high quality chairs. A common reaction is "They don't look at all orthopaedic."

SOFAS

Everything I've said about lounge chairs applies to sofas, probably more so. I've witnessed on a number of occasions the procedures people adopt when buying one. They pick out one they like the look of, walk over and sit down on the sofa, whose seat is usually far too low. They then fall back and wallow in its softness, which admittedly may initially appear the height of comfort. The Oxford English Dictionary defines comfort/ a state of physical ease and freedom from pain or constraint. This definition would apply to me for probably something in the region of ten to fifteen minutes, then the "freedom from pain" would disappear.

From experience I know the average depth of seat is eighteen and a half inches from the back of the knee to the back of the chair. The longest one I have ever made was twenty-three inches. The depth of sofa seats range from twenty inches to twenty-six inches.

To design and manufacture a custom tailored chair is comparatively easy, we've been doing this for twenty-five years. Sofas present a

Are you a slouch potato?

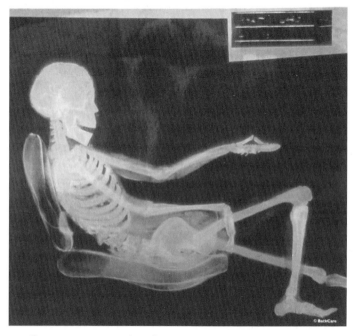

Support your back to maintain its natural curves. Remember to get up, stretch and move around every so often.

Contact: BackCare, 16 Elmtree Road, Teddington,
Middlesex TW11 8ST. Phone: 020 8977 5474
e-mail: back pain@compuserve.com website: www.backpain.org
BackCare, the new name for the National Back Pain Association,
is a Registered Charity (No. 256751)

**The Back Shop is a corporate member of BackCare
(The U.K. National Back Pain Association)**

problem, because they can be used by a number of people. We have overcome this by taking a large number of spinal profiles we have measured over the years and designing an average profile that would be representative of an average back. The backs of our sofas are made in this shape and give good support to most people.

I know very little about the aesthetic look of furniture. My son, Sean, who runs the business has an eye for furniture design and is forever bringing out new creations. Spinal System-S can be applied to any design. Should a customer want their own design of lounge chair or sofa it is simply a matter of giving Sean a sketch or a photograph and he will translate it into the customer's dream chair or sofa providing pain free sitting.

I have now gone through one day in my life and at no time did my spine move out of the "S." I suggest you look at what you did throughout the day before going to bed. Retrace your movements from lying in bed, getting out of bed and continuing on throughout the day as I did. Write an "x" on a piece of paper for every time your spine bent out of position, even if this was only for a split second. At "day's end" add up the total. Do the same for to-morrow. But this time make a conscious effort to reduce the number of x's. Continue doing this for a number of days. If you are conscientious about it you will find a big reduction. Then concentrate on the x's that keep reoccurring. Concentrate on one at a time. Find a way to eliminate it. Don't just consider the problem for a couple of minutes then decide there's no way around it. There are blank pages provided at the end of this book to facilitate this exercise.

When I am trying to find a solution to an important problem, I often apply "way out," or farcical, solutions to that problem. Quite often one of these fanciful solutions will trigger my mind into coming up with a practical solution I would never have thought of.

An illustration of what I am attempting to get you to do is to take any object at random, a paperclip, a shoe, a watermelon, a potato, etc.

Write down in thirty seconds the number of things of things you can do with one of these objects, say shoe. You would probably get half a dozen. The number I would get would be governed by the speed at which I can write. I would approach the game by looking around the room, as I am doing now and write down the names of every object in the room. So here goes: mouse, monitor, book, dog, telephone, television, window, ceiling, photocopier. etc. I approach the game by saying there is nothing I can't do with a shoe. Take the objects I have just mentioned, the sole of a shoe could double as mouse pad, the monitor a shelf to put the shoe on, the shoe could be one of a pair of book ends, to silence a mobile phone put it inside the shoe, the next is dead easy-what dog doesn't like playing with a shoe, television-the same use as the monitor or, if there is something annoying on, the shoe could be used as brick, a shoe is a handy object to keep a window open, stick the shoe on the ceiling as a piece of artwork, photocopy the sole of the shoe and so on.

You are probably asking what the hell has this to do with an awkward "x" that you can't eliminate from the daily grind. I am merely trying to emphasize that there can be many unexpected solutions to a problem.

CHAPTER TEN

THE NON INVASIVE REMOVAL OF A SLIPPED DISC

"The disc is not an essential part of one's anatomy"

I have already mentioned Alexander Walker-Naddell and his views on back surgery. Walker-Naddell, who died some years ago, was born and lived in Glasgow and was a Fellow of the Royal College of Surgeons. He specialized in orthopaedic and neuro-surgery and spent many years researching the anatomy of the spinal column.

He performed many operations on slipped discs and, as he writes in his book (The Slipped Disc and the Aching Back of Man):

"The results following this type of surgery, in my opinion, are not good, regardless of the surgeon who carries them out. The results were so variable that I decided to carry out some research in order to evolve a safer and perhaps more reliable method of treating a slipped disc. The eventual result of my researches was to produce a method of treatment whereby one could detach the offending disc without the use of surgery with all its attendant risks."

He carried out this research in the Pathology Department of Glasgow University:

"Dissecting the spinal column, particularly the area between the vertebral bodies. I noted that quite frequently the spinal discs were not present. These cases had died of various conditions but I took a careful history of these individuals from close relatives and found that 90% had no history of back trouble. This confirmed that the disc is not an essential part of one's anatomy and that the annular ligament which surrounds it is itself the spinal buffer or cushion. Thus the spinal column appears to function perfectly well when the disc is detached. As long as the annular ligament is present there will be no narrowing of the space between opposing vertebral bodies and thus supported by this ligament, they will not collapse on each other."

Walker-Naddell describes his Treatment for the Non-Surgical Detachment of a Slipped Disc Under Local Anaesthesia:

"During the hours spent dissecting specimens in the Pathology Laboratory I explored in every detail the intricate mechanism of the spinal column and realized that, in order to treat a patient, I could work "blind," as it were, with my fingers to detach the disc by pressing it against the sharp cutting edges of the torn annular ligament. I also soon realised that, in order to gain access to the site of the disc during treatment of a patient, the muscles in spasm, because of compression of the nerve root, must be made to relax. This could be done simply by the injection of a local anaesthetic into the nerve root affected. With the patient's muscles, previously in spasm, now completely flaccid, the spinal column could be flexed to an appreciable degree to allow access for my thumb to the site of the slipped disc.

This process, developed from these years of research in the Pathology Laboratory, I call the non-surgical detachment of a slipped disc under local anaesthesia or nerve root block. It is the essential and unique part of the treatment, for it provides a permanent cure for the slipped disc condition and yet involves none of the attendant risks of surgery.

The patient is asked to keep mobile after this treatment and, in fact, to adopt a training programme prescribed by me and varying according to the position of the disc lesion. This gentle activity will encourage more disc to protrude through the tear and thus at each subsequent treatment (normally 3 or 4 in all) all the disc will eventually be detached."

Walker-Naddell set up private practice some 30 years ago and achieved so much success with his unique treatment that his reputation spread, not only in the UK, but also throughout the world. I first heard of him some ten years ago when my long time friend Frank McCabe, who was then living in the States, visited him for treatment that was completely successful. Frank describes him as a brilliant evidence based surgeon who was extraordinarily positive about the outcome of his procedures, which were consistently successful. Unfortunately he did not peer review his research, apparently because his contemporaries downplayed his breakthrough.

In spite of this it has always been a source of some amazement to me that other medical people, who were aware of his success, have not tried to carry on his work.

Before publication of this book I gave the manuscript to one of the best-known medical experts working in the area of back pain relief. After six weeks when I hadn't heard from him I phoned to learn he hadn't read it. Two weeks later he phoned me. "I agree with everything you've written," were his opening remarks. To say I was surprised is greatly understating my reaction. It has been my experience that medical consultants tend to be aloof, condescending and dismissive of anything not in keeping with rigid medical practice.

He also said, "I don't think you should give people so much information about the spine." I accepted this criticism without reply. This was so fundamentally opposed to everything I have been saying over the last twenty-five years that I wisely said nothing. Anything else would have resulted in a heated rebuttal on my part and anyway I asked for a critical reaction and was getting it. Was he motivated by the old proverb, "A little knowledge is a dangerous thing?" Or was it something deeper?

When I asked him what he thought about Walker-Naddell's work, he replied " I would not read anything unless it was published in The Lancet," was his pompous reply. The Lancet is the authoritative voice of the medical profession.

"Where would the world be now if engineers took this attitude," I snapped back.

Walker-Naddell was a pioneer who, dissatisfied with the high failure rate of surgery, combined his skill as a surgeon with the detailed knowledge of the spine he acquired from his experimental work in the pathology laboratory. He invented a better way for detaching a ruptured disc. He must have had tremendous grit and determination to fly in the face of conventional medicine. These characteristics and what must have been a high degree of manual dexterity would be extremely difficulty to come by and so his work may never be replicated.

Perhaps his work may motivate others to invent a different technique for the non-surgical removal of a ruptured disc.

If the spine were an engineering structure surrounded by a casing and I was asked to consider a way of removing a ruptured disc without damaging the casing, I would give great consideration to the fact that the extruded disc is jelly-like and when young has a ninety percent water content. I would examine the possibility of inserting a syringe through the casing and into the jelly, then sucking the jelly out through the needle of the syringe. This is such an obvious solution that it must have been tried and found wanting. Perhaps the disc is too viscous to be sucked by conventional means. I might experiment with different suction techniques, try a range of vacuums, maybe introduce a solvent solution to reduce the viscosity of the disc to make it amenable to suction and so on.

I have researched these two methods to discover they have been tried and found wanting. As the moisture content of the disc decreases with age it becomes more viscous and so is immune to suction. A solvent would reduce the viscosity; however a side effect would be damage to the annular ligament.

Walker-Naddell was able "to detach the disc by pressing it against the sharp cutting edges of the annular ligament." Again if the spine was an engineering structure, I might consider how to detach the disc by mechanical means. I would look at the possibility of laying the structure on a flat surface, supporting it in the "S" and then applying increasing rates of vibration in the hope of sufficiently agitating the disc so that the on-off contact with the sharp edges of the tear in the annular ligament would result in it being detached. If the disc were frozen it would become stiff and would be easier to detach.

Another method might be to explore an advanced form of manipulation. Could the disc complex be manipulated in such a way that the tear in the annular ligament be opened and closed so that sharp edges would act like a scissors and cut the protruding disc?

These are just some thoughts on how Walker-Naddell's pioneering work may yet be of lasting benefit to back sufferers. It is my opinion that somewhere in the above hypotheses lies the germ of an idea, which could eventually revolutionize the treatment of back pain. It needs somebody with vision, initiative and perseverance to perfect the "Treatment for the non-invasive detachment of a slipped disc."

If I were asked to propose the most likely treatment for the non-surgical removal of a slipped disc I would give consideration to designing a system based on a syringe and using an inert gas at below freezing that would allow the 90% water content prolapsed disc to be frozen. Operating the same equipment and using the same type of inert gas, now at a temperature of around body heat, I would direct a fine jet of gas at the prolapsed disc. I would slowly melt the prolapsed disc, starting near the point of contact with the nerve root and working back to the annular ligament. The cartilage plates on the surfaces of the vertebrae provide the disc with nourishment. When the disc is detached from its nutrient supply it quickly starts to atrophy (waste away) and causes no adverse symptoms.

This is an engineering approach and may not be feasible from the medical perspective. For instance would the introduction of a foreign body, an inert gas, be harmful? Would the body absorb it like a prolapsed disc? Would freezing be harmful? And so on. If nothing else these hypotheses may motivate research into a completely safe method for removing ruptured discs.

However, thinking along these lines is to completely ignore the fact that a very successful solution exists and is pandering to medical obstinacy and negativity. The fact is that Walker-Naddell, a Fellow of the Royal College of Surgeons, has proven his non-invasive technique over a period of thirty years. He offended the medical profession by not conforming to medical practice. So what! He was a "one off," an innovative pioneer, who did his own thing. It is regrettable that politics should stand in the way of his work being used in combating the second most common pain.

CHAPTER ELEVEN

THE SIGNPOST AT THE CROSSROADS

"There are three signs saying "Do Nothing,"
"Invasive" "Non Invasive." Which do you take?

The way people react to an outbreak of back pain varies from person to person. I am discounting pain due to injury that requires immediate medical treatment. The pain may be a first time experience, or it may be the elevation of niggling pain to a higher level. I know of no scientific study, however it is my opinion that the majority, almost as a reflex action, seeks help from a G.P., who is a non-specialist in spinal disorders. I have mentioned earlier about a lecture I attended given by a doctor attached to a large UK insurance company who, in relation to the sudden outbreak of back pain among workers threatened with redundancy, said that if a worker complains of a back pain there is no known method to disprove this claim. In every case a deal has to be done. This statement will give you some idea of the extreme difficulty involved in diagnosing the cause of back pain.

"Determining the cause of a given individual's pain, however, often remains more art than science."(Scientific American 1998). And this is using the most up-to-date technology.

The National Back Pain Association (BackCare) in their book "Back and Neck Pain" state, **"There are about fifty bones, one hundred joints, one thousand muscles and a million nerves in your back and neck. So it is hardly surprising that back and neck pain are among the most common of physical afflictions. There is so much to go wrong,"** and **"The enormous complexity of your back**

**and neck probably makes it impossible to identify specific tissues
as the cause of your pain."**

So what is the non-specialist overworked GP, with no diagnostic
equipment, to do? That's right, he kicks to touch, by prescribing
painkillers. Many customers tell me this is accompanied by remarks
such as, "wear and tear," "a touch of arthritis," and one I have already
mentioned, "cervical spondylosis."

When I was carted off to hospital twenty-five years ago I vividly
remember some pills being pushed into my mouth and being told to
swallow. This procedure was repeated on the second day of my stay.
On the third day they were either overworked, or thought I was on
the mend, because the pills were left with a glass of water on my
bedside locker. I promptly pushed the pills under my pillow before
later hiding them in my toilet bag. Whether my squirreling was
noticed, or not, I don't know, but on one occasion an extra officious
nurse stood overseeing the popping in of the pills, followed by the
gulping of water. The water went the intended route, the pills being
removed once the "authority" was out of sight.

Writing this reminds were of how so utterly dependent, vulnerable
and subservient even the most independent of people become when
subjected to hospital authority. They become non-people.

Being reduced to a non-person didn't interfere with my logical mind.
If I took the pills I would not have pain. That would mean that if
asked once more to touch my toes I could make the effort without
pain. (Since I was sixteen the furthest I could bend was to a point just
fractionally below my knees.) And I probably would have made the
effort as I was demoralized. But I knew this would cause more damage
to my spine. I also deduced that the reason I had pain was that my
body was telling my brain that a fault existed and that the area of the
fault was to be protected to prevent more damage. So I didn't take the
pills then, or afterwards.

GP's prescribe painkillers, pain is then either eased or eliminated and
the continuation of the routine of daily spinal abuse compounds the

problem. When the course of painkillers is at an end the pain may return. This is followed by another visit to the doctor for more painkillers. When it becomes obvious that the prescribed treatment is not working the patient is referred to a medical specialist and so a cross roads on the way to possible recovery has been reached.

The signpost at the crossroads has three signs.

The one pointing straight ahead reads, **"Do nothing."** In smaller lettering underneath and in brackets, are the words **"(at your peril.)"**

I would have gone straight through this crossroads many times in my early, middle and late twenties and early thirties. I wouldn't have seen the crossroads, so how could I have been expected to see the signs. I knew nothing about the spine and, apart from being aware of the original injury, gave absolutely no thought to the reason for my regular bouts of pain. Eventually the inevitable happened, my disc ruptured and my life changed forever.

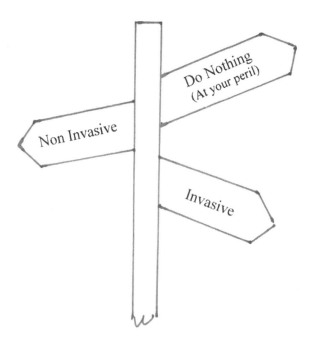

My hope is that having read what I have written so far you have a good idea of the structure of the spine, the cause of back pain and its prevention. This awareness will enable you to see the crossroads the signpost and the possible disastrous consequences of going straight ahead and continuing to abuse your spine on a constant basis.

The sign pointing to the right reads **"Invasive."** (The Oxford English Dictionary defines; **invasive /** involving the introduction of instruments into the body.) Taking this route could ultimately result in invasive surgery. And remember, as already mentioned, Professor Deyo, Dept. of Medicine, University of Washington states;

"Recent studies show that even for patients with a herniated disc, spontaneous recovery is the rule."

and

"The actual proportion of all back pain patients who are surgical candidates is only about 2 percent."

The **"Invasive"** sign is large containing many cautionary notes, you are already aware of these:

"It must be understood that this is major surgery and not to be undertaken lightly. Unfortunately, also, the results in most cases leave much to be desired." (Alexander Walker-Naddell FRCS)

"No matter how successful the operation is, your back will never be the same again; surgery creates scar tissue, which doesn't exist in a normal back." (Hamilton Hall, FRCSC)

As I have said a number of times we would have measured something in the region of ten thousand spines and have spoken to over thirty thousand other back sufferers. From this large group the number who had back operations would be in the region of one in ten, or four thousand. The number of successful back operations I have encountered is less than one percent. Admittedly this is an extremely imprecise finding, nevertheless it is frightening and warrants a properly conducted survey (I have already mentioned some parameters) to enable candidates to realize that there is less than one percent chance of a quick fix.

I have come across operations that were successful for a period of time, the longest being thirteen years, before the original pain returned. Many people have told me the surgeon told them eighty percent of operations are successful. They never think to ask what is meant by the word "successful." Most have absolutely no knowledge of the spine, cannot ask intelligent questions and meekly accept what they are told.

I know I've said this already however, because of its importance, I'll say it again:

I define a successful operation as one that completely and permanently restores the quality of living that existed prior to the outbreak of the pain that resulted in the invasive procedure.

Non-invasive treatment (physiotherapy, chiropractic, osteopathy, Spinal System S, exercising, etc.), combined with the most important one of always keeping the spine in S will in the majority of cases result in pain free living.

We know the famous Mayo Clinic in the States has a surprisingly frank notice on its website;

"Long term outcomes also are often similar following less invasive treatments."

However, based on my experience as an engineer working exclusively in the area of back pain relief for twenty-five years, I would disagree with the use of the word "similar." It is my absolute conviction that operations should only be carried out in a very small number of cases. The person who refuses one can live a pain free life provided help is sought and life style changed. If I can achieve this so can you.

I would interpret the above statement by the Mayo Clinic as an admission that most invasive operations are unnecessary and should not be carried out. If the body is capable of healing itself with the help of non invasive treatment nobody should agree to going under the knife until every last avenue has been fully explored and found wanting. It is my opinion that before undergoing elective surgery every patient should be required to give proof that non invasive treatment was performed and found to be unsuccessful.

"The more selective surgeons have strict guidelines and operate only if there is evidence of the nerves in the saddle area and legs not working properly." (Sarah Keys in Back Sufferers Bible)

On a personal basis, when my daughter Tara was in hospital there was a total of six patients in her ward. On average two were either awaiting back operations, or recuperating. Half way through her two-month stay her favourite nurse disappeared. I was delighted to meet her in the corridor two weeks later; Tara was upset by her absence. I asked her where she'd been.

"I had a back problem," she replied.

"What did you do about it?" I asked.

"I went to a chiropractor."

"Why didn't you come here?"

She smiled. I smiled. We both went out separate ways.

The sign pointing to the left reads **"Non-Invasive."** Underneath is listed a large number of alternative treatments that can successfully treat back pain.

But one of the main points I have been making is that you cannot hand over the solution to your back problem to third parties. They have a great deal of success in treating the symptoms, but are powerless to eliminate the causes. Only you can do this. I am often asked which treatment, chiropractic, physiotherapy, osteopathy, acupuncture etc, is the best. It has been my experience over the last twenty-five years that each of these professions is very successful in treating back pain and after a number of sessions patients experience relief. When the course of treatment is completed patients are often given a list of daily exercises aimed at keeping their backs in good trim. Some patients carry out these exercises, others don't.

The latter fall precisely into the category of those who hand over the solution to their problems to a third party, will not take responsibility for their own bodies and do nothing to help themselves.

The former are also guilty. Of course their intentions are well meaning as they conscientiously do the prescribed daily strengthening exercises designed to keep their spines in good condition. However, once daily exercises are completed the majority then forget about their spine. Off they go to work in the car, bending over a desk for hour after hour, abusing their spines on a constant basis, with no thought to "S." The fact that they do exercise will probably result in a longer interval of time before the next outbreak of more pain, when compared with the non-exercisers.

Take great care if you are exercising that it does not involve bending the spine forward. Throughout my sporting life I constantly came across the command, "Bend, touch your toes!" I could never do this and always wondered what the objective was. I have seen teammates alongside me being able to put the palms of their hands on the ground and yet they didn't have the same level of fitness that I had. You will remember the surgeon commanding me to this when I was in hospital awaiting an operation and the physiological effect it triggered. So be careful.

The ideal way to go about your day-to-day life is to do daily exercise in combination with keeping the spine in the "S." I don't do any structured exercise. I am mostly on my feet and moving around during the day, have two dogs and so do a lot of walking, play golf, garden, am constantly doing jobs around the house and so keep fairly fit. Last weekend for example I spent six hours repairing the roof of my house. This meant kneeling down repairing the fault (I always use kneeling pads), umpteen trips up and down a ladder fetching tools and equipment, even lying on my side to fit roofing felt under the roof tiles. The next day I spent a similar amount of time cutting a hedge. As the hedge is about ten feet high, this meant more climbing up and down a ladder and kneeling to pick up the leaves. So it works for me and as I've said before, "If something works, stay with it."

So the question you must ask yourself now is do I go to the left, or to the right.

I was too scared to take the road to the right and so went left. However I ignored all the existing signs and made one of my own which read **"S,"** and it proved to be successful. I certainly didn't know when I made my decision that,

"Studies using repeated MRI revealed that the herniated part of the disc often shrinks naturally over time and about 90 percent of patients will experience gradual improvement over a period of six weeks." (Professor Deyo)

So I was lucky in this regard. I knew sitting in the C position was the cause of my pain. So I sat in S. At all times throughout the day, no matter what my activity, I keep this S position. I am now pain free.

The following is a list of some non-invasive treatments. It is up to you to choose the one that you think is most suitable to your requirements.

Acupuncture
Acupuncture helps back pain by reducing muscle tension and so relieves pain and improves mobility. This is achieved by inserting fine sterilized needles into the body. The technique is based on the ancient Chinese art of acupuncture.

Chiropractic
Chiropractors believe that most back pain is caused by "subluxations of the various vertebrae" that causes pressure on nerves, with

resultant pain. (The Oxford English Dictionary defines "subluxation" as "partial dislocation.") Their objective is to heal the whole person along with the specific problem.

They use spinal manipulation and also specially designed couches, divided into four sections, which can be lowered or raised independently of each other. They manipulate the spine by thrusting down sharply on the offending area of the spine. The section of the couch supporting this area immediately collapses, resulting in instant force being applied to the vertebrae to be manipulated.

McKenzie Technique

The McKenzie approach is based on the principal of teaching patients to treat themselves through prescribed self-generated movements. By monitoring the response of low back and neck pain during a mechanical spinal assessment it is possible to design an individualized program of exercises for each patient.

Manipulation

Many practitioners specializing in back pain relief carry out manipulation. The usual procedure is to first massage the back to relax the muscles. The spine is then manipulated using different methods, depending on the practitioner. The objective is, by extending and rotating the spine, to move whatever is pressing on the nerve, a bulging disc for example, out of contact with the nerve and so eliminate the pain.

My falling down the stairs in Fitzwilliam Square some twenty years was an extreme example of manipulation. However, as the television show showing dare devil tricks announces at the end of each episode, "Do not attempt to try this at home."

Osteopathy

Osteopaths contend that the human body has the power to combat disease and is similar to any piece of mechanical equipment that is liable to malfunction due to wear-and-tear and breakdowns. The cause of disease is dislocated bones, disorders of ligaments and cramped muscles, putting pressure on the circulation and nerves. This leads to lack of blood supply to the relevant tissues and the death of tissues, partly due to the blocking of the life force that moves around the nerves in the body.

While most people attend osteopaths to get pain relief, the object of osteopathy is to get the complete body in perfect mechanical order. The belief is that if they can correct problems in the skeleton, muscles and nerves, this will help the body to heal itself, even problems in the body which are some distance from the spine. They achieve this by gentle manipulation of the spine. Their techniques are more concerned with loosening and freeing rather than repositioning

Pilates
Pilates is a gentle form of exercise that can be practised by people of all ages and physical capabilities.

Physiotherapy
Physiotherapists examine patients to make a diagnosis and determine treatments. These treatments include exercise, mobilization and manipulation, electrical ultrasound and heat treatment.

The hands are the physiotherapist's main tools. Massage is used as a relaxant prior to manipulation, which consists of gentle rhythmic movement of the spinal joint.

Reflexology
The objective of reflexology is to eliminate waste from the body, to improve circulation and to bring the body into balance. Reflexology is based on the fact that every organ in the body has a mirror image in the feet and hands. The area in the foot where the spine is represented is from the side of the big toe to roughly half way down the instep. By running the thumb lightly over this area the reflexologist can detect the twenty-four tiny nodules representing

every vertebrae. By lightly pressing each nodule with the thumb the reflexologist can detect the nodule that is the mirror image of the vertebra at the seat of the pain. (Pressing this nodule causes some discomfort, pressing the other twenty-four causes no reaction.)

The reflexologist, by gently massaging the nodule, is able to bring about pain relief. There are great many things in life that I don't understand and this is one of them. But it works. I have personally witnessed a patient limping into a reflexologist in great pain, having a treatment session and then leaving pain free without a limp.

CHAPTER TWELVE

THINK S

"Spinal System-Life - for pain free living"

Having read all the foregoing you are aware of how I came to be pain free. It took a huge amount of hard work, research and determination to get to this condition. If you want to adopt my Spinal System-Life with the objective of being pain free, you must resign yourself to the fact that it will take a great deal of discipline; not in a haphazard fashion, but on a continuous twenty four hours per day basis, not for a month, or a year or two, but forever. The good thing is that with the passage of time your body and mind adjust to the new constraints being imposed and "Spinal System-Life" becomes second nature, eventually requiring little thought or effort.

The danger period will be when you are eventually pain free, become complacent and slip back into your old bad "C" habits.

You may continue to luxuriate in a wellbeing sense of pain free contentment for some time, but in all probability the pain will return. And it can strike suddenly; brought on by a cough, a stumble, stepping off a footpath and any number of minor accidents.

In fact there doesn't have to be any noticeable cause. You can wake up one morning and "bang" the agony is, dare I say, back.

"Diagnosing the cause of back pain is more an art than a science," so too is trying to figure out why it has returned.

In summation:

To achieve pain free living Spinal System-Life demands you obey three simple rules.

---

1. You cannot hand over the solution to your back problem to a third party.

You can seek help to relieve the pain. In most cases this is successful. However sooner or later the pain returns, resulting in more treatment. The reason is the symptoms and not the causes are being treated. Only you can eliminate the causes.

---

2. If you keep your spine in its own unique S shape at all times in the vast majority of cases you will be pain free.

Our ancestors did this, as do primitive tribes, resulting in low incidences of back pain. We, on the other hand do not and as a consequence eighty percent of us suffer back pain at some time in our lives.

---

3. You must stop arranging your body around the work.

Work can be defined as any type of activity; housework, gardening, DIY, office work, watching television etc.

If you are not a back sufferer you are now better equipped to prevent becoming one.

If you are either of the above: think S.

Think S

Every day throughout the day

Driving, at work, at play.

Particularly sitting.

Don't arrange your body

Around the work

But vice versa.

Seek help and advice.

Use it with S.

In most cases

You will be pain free.

If not

C may not be the cause.

But it will be comforting to know

Disc pressure will be the lowest possible.

CHAPTER THIRTEEN

HOPE

"Hope that in the future if he carried out my suggestions he could be pain free"

On a daily basis I meet back sufferers who are in great pain. Many are deeply depressed, having been told nothing more can be done for them. They have been dismissed with callous advice such as "get on with your life as best you can." Left without hope they deteriorate into states of utter despair and having no knowledge of spinal anatomy they compound the problem by continuing to abuse their spines.

It has been my experience that talking to these people, offering hope by explaining how the spine works and how to live with the least possible pressure in the inetevertebral discs, can result in an immediate change in their outlook. Their meek acceptance of the opinion of an expert that their back pain is there forever is replaced by a steely resolve to take responsibility for their own problem and to set about solving it.

The best example of this happened the Christmas before last.

It has always been my custom to return to work on my own around the 29th December and did so on this occasion. At 10 a.m. the street outside The Back Shop, situated on Exchequer Street and about two hundred yards from Dublin's premier street, Grafton Street, was totally devoid of people. So quiet was the city that as I switched on heat and lights it crossed my mind that perhaps it would have been advisable to have waited another day before opening up.

On the point of taking off my overcoat, the door squeaked open and a stocky figure, wearing a tightly fitting corduroy jacket and baseball cap with an oversized peak, shuffled painfully into the shop. Head bent forward, as if unsure of his footing and examining the ground ahead for obstacles, he walked slowly towards me. Arms hanging by his sides, every muscle in his body completely rigid, he didn't bend his knees, just stuck one rigid leg out in front of the other. I walked to meet him and stooped slightly to look under the peak of the baseball cap. A pair of slitty eyes set into a waxen face, completely devoid of expression, met my gaze. I estimated his age to be about sixty five. Neither of us spoke for some time. I took up the running.

"You're in pain!"

He didn't answer, just a slight nod.

"Would you prefer to sit or stand?"

"Stand," he had broken his silence.

With a voice barely rising above a whisper he started to tell me about himself. He had been a professional soccer player and had an operation to fuse two discs in his lumbar spine. The operation was unsuccessful and he had to retire. He had managed to survive the fifteen years since the operation by having an epidural injection every three months. The injection relieved the pain for about two weeks and throughout the next ten weeks he would suffer the most excruciating pain until another injection would give temporary relief. Telling me this I could see he was now at the end of his tether, in deep despair.

I asked whom he played for and he mentioned two well known English First Division clubs. There was something about him that was vaguely familiar and when he told me his name the jigsaw clicked into position. I clearly remember him as a wholehearted defender, the type that journalists would describe as willing to go through a brick wall and who received in the region of thirty caps playing for Ireland. In extreme pain and home to visit his mother for Christmas, somebody had told him about The Back Shop. He was returning to England the next day to face another operation on his back.

He knew absolutely nothing about the spine, the causes of back pain and how back pain can be prevented. As I spoke to him customers started to come into the shop at frequent intervals. He told me he wasn't in a hurry and listened as I spoke to each one about their particular problem. At one stage there were fourteen people, including him in a round table discussion. (By this time I had him sitting on a car seat fitted with one of my custom tailored car seat moulds.)

He was with me for probably an hour and a half and during this period I measured him for a car seat mould. His job entails a lot of driving and I promised to post the car seat mould to him in the second week in January. I also gave him a Posturite lumbar support with instructions to never again sit without this support at his lumbar spine.

Then an extraordinary thing happened. He shook my hand, said goodbye, turned on his heel and walked with free flowing strides to the door. Yes. Walked. He was out the door before I saw the significance. I strode after him. He heard me coming, turned to meet me. His eyes weren't slitty anymore. They were wide open and smiling, as was his face, now no longer waxen but a healthy hue. I could see that he wasn't sixty-five but in his middle forties.

I can't say for certain what brought about this transformation. Maybe it was sitting in the car seat mould, or having his spine pushed into perfect posture in the measuring chair. But if I had to bet on one reason it would be that I had given him hope. Hope that another operation could be avoided; hope that in the future if he carried out my suggestions he could be pain free.

Having read "The Engineering Solution to Suffering Back Pain" it is my hope that it will have the same affect on you.

If you put into practice what you have read I would be grateful if you could inform me of the consequences. It would of great benefit to have a record of back sufferers who have taken responsibility for their own problem, have done something about it and are now pain free.

CHAPTER FOURTEEN

THE FUTURE

"Replacing them by signals denoting the spine is still unstable"

I said previously, "The spinal column is like an ingenious piece of mechanical engineering design controlled by a computer so sophisticated that it may be centuries before man is capable of fully understanding how it works." I said, "It appeared as if it was designed to cope with a way of life that existed many thousands of years ago, when man never sat." I also said that if man had designed it it could be modified to take into account present conditions. But it wasn't and it can't.

What I am proposing in this book is a method to help you overcome the effects of subjecting the spine to a condition it would appear it was not designed for; the sitting position. I admit the method means accepting the philosophy of what I propose, which is a huge hurdle in itself as many people can be sceptical. Also it requires high levels of dedication, determination and constant awareness of the fact that even a small deviation can have painful consequences.

Most people want a quick fix, to cure the problem by handing it over to a third person and as you now know my first rule for pain free living says this is not feasible.

But what about the future? A hundred years ago many thousands of things that we now take for granted were not even contemplated; mobile phones, computers, space travel, heart transplants, I am now reading about transplanting of whole faces and so on.. So why draw the line at overcoming the fact that the spine was not designed for sitting? If I were a young engineer starting off in my profession and wanted to devote my life to one major project I would choose this one. I would give great consideration to this design flaw and how to overcome it to benefit the whole of mankind. You cannot alter the spine itself, but what about the software that controls its movements?

We already have a very good clue as to a possible solution. I have spoken about what happens when walking, on a footpath for instance and you stub your toe resulting in you losing balance and starting to fall. In most cases you don't fall. Close your eyes and visualize. You are falling forward. Then suddenly the fall is arrested. Without any conscious thought or action on your part, your body is restored to the vertical. You may feel a bit shocked for a couple of seconds, but you continue your walk as if nothing had happened. I would suspect that most people would have given this phenomenon little thought. I think it could play a huge part in man being able to design a system that could result in the spine being able to fully cope with modern living conditions.

It would always be in S, unless commanded by the brain to move out of position.

The human body has an in-built balancing system, located in the inner ears, that is constantly working to keep the body upright; riding a bike is a perfect example. You are totally unaware this is happening, until an extreme condition occurs like stubbing your toe when, without any conscious effort on your part, you are restored to the vertical.

So we already have a system that restores the body to the vertical when involuntary movements occur. What happens is that the balancing system detects an out of balance and sends electrical signals to the brain by way of the balance nerve. The brain interprets this signal as head movement and commands the large muscles in the spine to pull down on one of the three bony outgrowths (two transverse processes sticking out one to the left the other to the right and the posterior process sticking out backwards) at the back of each vertebra until balance has been restored. Falling forward results in the muscles pulling the posterior processes downwards. Falling to the side results in the muscles pulling the appropriate transverse processes downwards. This doesn't happen when you sit. The system described above goes into repose. Without special techniques, such as building up the stomach muscles, your spine slumps into a forward C position, which as you know will eventually result in pain.

The solution to the future prevention of back pain, as I see it, is to devise a system that would allow the balancing system to operate involuntarily, not only in the dynamic position, but also in the seated position.

A great deal of research would be necessary. This would concentrate on what exactly happens when the body suddenly moves without conscious effort off the vertical, resulting in a tiny electrical signal alerting the brain to this fact and remedial action being taken.

The balancing system in the inner ear consists of minute chambers containing fluid and delicate nerve endings. As the head moves, the fluid in the balancing chambers moves and makes contact with the balance nerve endings. The nerve endings detect the change and send electrical signals to the brain by way of the balance nerve.

A complex interaction of four body systems; the inner ears, the eyes, the brain and the spinal cord along with joints and muscles of the limbs maintain a person's sense of balance.

Having a complete understanding of this dynamic balancing system could provide the solution to balancing the body in the seated position. It would then be a matter of reprogramming the computer, the brain, to always pull the muscles in the spine taut in the seated position and so keeping the spine in S. This would be the fail-safe position requiring a conscious thought to instruct the spine to bend forward.

It would be necessary to analyse why in the seated position the brain switches off the dynamic balancing system. By making use of this information is there some way the brain could be duped into believing the dynamic balancing system is still in operating? Could a microchip be inserted in the electrical circuit between the balancing system in the ears and the brain that would block out commands announcing the person is now seated and replacing them by signals denoting the spine is still unstable. The spine would then always be in S while seated and only a conscious command would override this.

This may appear fanciful. But is it any more so than it being suggested sixty years ago that in the future you could pick up a mobile phone, dial a number and within a couple of seconds you could be speaking to somebody as far away as Australia? Use the phone as a camera? Send the photograph to the same person in Australia? Within seconds a woman could be receiving images of her newborn grandchild.

Compared with the remarkable advances in technology over the last couple of decades, maybe my proposition of adapting an involuntary action of the body to preserve its equilibrium in one position and applying it to another position to preserve S is not that outlandish.

CHAPTER FIFTEEN

SPINAL INFORMATION

The following is a more comprehensive list of some terms used in connection with the spine:

Alternative Treatment
Non surgical treatment of back pain. This is a large subject ranging from chiropractic, physiotherapy, osteopathy, to exercise, manipulation, massage, bone setters, pilates, yoga, Spinal System-S etc.

Ankylosing Spondylitis
A disease of unknown origin that causes the intervertebral disc complexes, and ligaments, to stiffen over a period of time. Eventually they become rigid, resulting in the spine becoming completely inflexible. It extremely important that proper treatment include instruction in postural habits, so that as the spine slowly fuses it will be fused in its optimum position, the "S."

Annular Ligament
The tough layers of fibrous tissue that bind the vertebrae together, and enclose the soft jelly-like disc (nucleus pulposus).

Archnoiditis
Dense scar tissue that forms around nerves leaving the spinal canal after an operation, resulting in pain, pins and needles and numbness.

Cerebrospinal Fluid
The clear fluid that surrounds the brain and the spinal cord.

Cervical
This means neck. There are seven cervical vertebrae, designated from the top down as C1, C2 etc. to C7. They are small in comparison with the lumbar vertebrae.

Coccyx
The last bone in the spine, composed from four tiny fused vertebrae, supposedly all that's left of a tail we inherited from the apes millions of years ago.

Cat Scan (Computerised Tomography)
A system that produces cross sectional images from information received through x-rays going through the body.

Congenital
Being present at birth.

Degeneration
A process of change that takes place, normally with age, in bone or soft tissue.

Degenerative Disc Disease
The intervertebral disc system changes from a supple, flexible, structure designed to absorb forces in the spine, caused by daily living, to a stiffer more inflexible one that restricts movement, and is less able to cushion these forces. This occurs naturally with age as the ninety percent moisture content of the disc system gradually reduces. The process can start as early as the twenties and by the 60th year it would be a normal occurrence.

Disc (Nucleus Pulposus)
One of 24 jelly-like sacs that form one third of the intervertebral disc system.

Discectomy
The surgical removal of part of the soft jelly-like disc that has ruptured (herniated, or prolapsed) and has been extruded through a tear in the annular ligament that encloses it.

Epidural
Epidural injections, given to relieve sciatica (leg pain), contain a local anaesthetic and corticosteroid, and are injected into the

epidural space between the bone and the membrane that encloses the spinal cord.

Facet Joints
The spinal vertebrae are attached together by the disc complex. At the rear of the body of each vertebra, and part of it, is the vertebral arch that has a hole (the spinal canal), three finger-like bones protruding downwards and backwards (with muscles attached these act as levers for moving the spine into different positions), and four bony pieces, two on top of the vertebral arch and two below. These are the facet joints. They interlock with two corresponding facet joints on the vertebra above and the vertebra below. The area of bone where they lightly touch is covered in cartilage (like the knuckle in a chicken bone) to allow fleeting contact. Facet joints guide the spine when moving up and down, and provide a limitation to twisting and bending movements. The facet joints are enclosed by tough elastic sheaths called capsules that knit the two sides of the joint together, and dampen all jolts passing through the facets.

Instability
Abnormal increase in the movement of one vertebra to another.

Intervertebral Disc System
Between every two vertebrae is the intervertebral disc system that consists of three components;

(1) The Annular Ligament. Composed of layers of tough fibrous material, and attached to the top and bottom vertebrae.

(2) The Disc. A soft jelly-like sac, 90% water content, and enclosed by the annular ligament.

(3) Two Cartilage Plates. Highly polished circular plates attached to the upper and lower vertebrae. Just as the annular ligament encloses the soft disc from side to side, the cartilage plates enclose it from top to bottom. The plates provide the disc with nutrient. The disc remains healthy as long as it is in contact with the plates. Break contact and it wastes away.

The function of the disc system is to allow spinal movement, and to cushion the forces that travel up the spine as a result of daily living, walking, running, etc. It functions something like a motorcar suspension system, with the coiled spring representing the annular ligament, and the hydraulic damper representing the disc.

Kyphosis
The convex curve of the thoracic spine. It can be over exaggerated in such diseases as osteoporosis.

Laminectomy
An operation which consists of breaking a hole in the vertebral arch at the back of a vertebra in order to gain access to the nerve roots leaving the spinal canal.

Locked Back
When a fault occurs in a joint a signal is sent to the brain, which instantaneously instructs the nearby muscles to contract, or "lock." The objective is to protect the joint from further damage.

Lordosis
The concave curve in the lumbar, and cervical spine.

Ligaments
Bands of fibrous tissue that bind joints.

Lumbar
The area of the spine between the thoracic vertebrae and the sacrum. There are five vertebrae in the lumbar spine numbered L1 to L5 downwards. The majority of back problems occur in this area.

MRI Scan
Magnetic Resonance Imaging is a sophisticated scanner that uses magnetic fields and computer technology to provide images of the internal anatomy of the body. This includes prolapsed discs and nerve

roots. However, as much back pain is not caused by ruptured discs (20% of <u>pain free</u> people under 60 have ruptured discs), the fact that an MRI tells you that you have a ruptured disc only proves you have a ruptured disc. It by no means tells you this is the cause of your pain.

Myelography
A water-soluble dye, impervious to x-rays, is injected into the cerebrospinal fluid surrounding the spinal cord and nerve roots. This allows nerve tissues to be viewed on x-ray.

Nerve Root
The point where the nerves leave the spinal cord before going right and left to serve an area of the body.

Osteoarthritis
Caused by degenerative change of a joint. As part of the aging process the intervertebral disc system gradually loses moisture content, resulting in adjoining discs coming closer together. This results in the facet joints, which normally only make fleeting contact with each other, making heavy contact, and the cartilage surfaces begin to grind against each other. Friction begins to roughen these smooth surfaces, nerves in the facet capsules surrounding the joint pick up the malfunction and convey it to the brain that instructs the surrounding muscles to contract to protect the joint. This causes inflammation of the joint by reducing the flow of blood through the capsule with resultant pain.

Osteophytes
Extra bone growing on the edges of vertebrae that can cause narrowing of the nerve canals, pinching the nerves, and causing pain.

Osteoporosis
Bone disease that decreases the calcium content of the bone in the vertebrae resulting in loss of bone density. This makes the bone liable to fracture and collapse, causing pain. With this condition it is

essential that the spine be always kept in the "S" in order to subject the vertebrae to the least possible stress and so combat fractures. It is more common in women after the menopause.

Perfect Posture
Every spine has its own unique "S" shape. This is perfect posture. When the spine is in the "S" there is the minimum possible pressure in the intervertebral disc system. Bad posture is when the spine moves out of the "S," with the pressure increasing as the spine bends forward until it reaches the "C" position, where it is greatest.

Sacrum
The base of the spine sits on the sacrum, a wedge shaped bony mass made up from five fused vertebrae.

Slipped Disc
This expression causes great confusion amongst the general public, a great number of whom have only a vague idea of its meaning. I know from meeting back sufferers on a daily basis that a large number of them are of the opinion that the vertebrae are kept apart by a semi-solid object they refer to as a "disc." Their favourite expression is "The chiropractor (or whatever expert they have been attending) pushed the disc back into position." The disc is of jelly-like consistency, has been squeezed out of position through a tear in the annular ligament that surrounds it, like toothpaste out of a tube, and like toothpaste cannot be put back in position.

Spinal System-Life
The system I devised and have rigidly followed for many years, that has resulted in me having a pain free, virtually unrestricted, lifestyle. (Spinal System-S on which I hold a patent is a major part of this overall system).

Spinal System-S
The engineering solution to suffering back pain when sitting.

Surgery
Medical procedure that involves cutting the body open to perform treatment.

Thoracic
The upper back, between the lumbar and cervical spine. There are twelve vertebrae in the thoracic spine, numbered from T1 downwards to T12. T1 is immediately below C7, T12 immediately above L1.

TENS (Transcutaneous Electrical Nerve Stimulation)
A machine that delivers small electric shocks through adhesively attached electrodes placed on either side of the spine. The objective is to block pain messages to the brain and to produce the body's natural pain killers, endorphins.

<u>NOTES</u>

<u>NOTES</u>

NOTES

NOTES

<u>NOTES</u>

<u>NOTES</u>

<u>NOTES</u>